THE SOCIAL HORIZON OF KNOWLEDGE

POZNAŃ STUDIES
IN THE PHILOSOPHY OF THE SCIENCES AND THE HUMANITIES

VOLUME 22

EDITORS

Jerzy Brzeziński
Andrzej Klawiter
Tomasz Maruszewski

Leszek Nowak (editor-in-chief)
Izabella Nowakowa
Ryszard Stachowski

ADVISORY COMMITTEE

Josef Agassi (Tel-Aviv)
Etienne Balibar (Paris)
Piotr Buczkowski (Poznań)
Mario Bunge (Montreal)
Robert S. Cohen (Boston)
Francesco Coniglione (Catania)
Andrzej Falkiewicz (Wrocław)
Ernest Gellner (Cambridge)
Jaakko Hintikka (Tallahassee)
Jerzy Kmita (Poznań)
Władysław Krajewski (Warszawa)

Krzysztof Łastowski (Poznań)
Theo A.F. Kuipers (Groningen)
Ilkka Niiniluoto (Helsinki)
Günter Patzig (Göttingen)
Marian Przełęcki (Warszawa)
Jan Such (Poznań)
Jerzy Topolski (Poznań)
Ryszard Wójcicki (Łódź)
Georg H. von Wright (Helsinki)
Zygmunt Ziembiński (Poznań)

The principal task of the book series "Poznań Studies in the Philosophy of the Sciences and the Humanities" is to promote the development of philosophy which would respect both the tradition of great philosophical ideas and the method of thinking introduced by analytical philosophy. Our aim is to contribute to practicing philosophy as deep as phenomenology or Marxism and as rationally justified as positivism or hypotheticism.

The address: prof. L. Nowak, Cybulskiego 13, 60-247 Poznań, Poland.

THE SOCIAL HORIZON
OF KNOWLEDGE

Edited by

Piotr Buczkowski

AMSTERDAM—ATLANTA, GA 1991

ISBN: 90-5183-270-2 (CIP)
©Editions Rodopi B.V., Amsterdam - Atlanta, GA 1991
Printed in The Netherlands

TABLE OF CONTENTS

*Poznań Studies in the Philosophy
of the Sciences and the Humanities
1991, Vol. 22, pp. 7–38*

Jürgen Ritsert

THE WITTGENSTEIN-PROBLEM IN SOCIOLOGY OR: THE "LINGUISTIC TURN" AS A PIROUETTE

SOCIOLOGY AND PHILOSOPHY are interrelated by colourful variety of dependencies and demarcation-practices. The normative philosopher of science on the philosophical side for example, wants very much to point out the way social scientists ought to take when developing methods, terms and theories. However, social scientists and historians of science follow this advice with a great deal of reluctance and take revenge by forcing sociological inputs into philosophy of science, so that philosophers are driven towards a Kuhnian view.

But philosophers can still laugh up their sleeves, for the great majority of social researchers — sometimes quite untouched by any philosophical zeal — have taken a linguistic turn in their theoretical evolution, a turn with a Wittgensteinian figure to it. It is this twist we will now examine.

It looks — and it is presented — as if this turn has saved social scientists from the ultimately fruitless efforts to develop (constitute) the world (objectivity) and the *alter ego* (sociality) from the self--consciousness of single reflecting subjects. It looks — and is presented — therefore, as if sociology had also managed to get rid of the Cartesian Ego as the measure of the real world and to replace it by making language the ultimate foothold for all attempts at knowledge. All this may turn out to be quite valuable as long as the last "remembrance of nature in the subject" [Horkheimer & Adorno, 1979, p. 40], of the chances of developing self and self-determination do not get lost.

In addition, we might well accept, with Wittgenstein, that a philosophical discipline determined to purify language could be of some assistance to sociology: "Philosophy is a battle against the bewitchment of our intelligence by means of language" [Wittgenstein, 1976, aphorism

109]. But I'm not aiming my criticism at these capacities of philosophy, analytical and otherwise. Rather I want to show that the "linguistic turn" is capable of causing mischief — at least for sociologists. This mischief stems from more than the language chosen for examination in any particular case.

Because there are so many taboos on research in "ontology" I happily chose the term "social ontology" to mark off the realm of problems I am interested in. They involve some answers to the question: what's holding society together — language for instance?

"Social ontology" marks the place where the troubles begin. I want to argue, that at the least, Wittgenstein's linguistic twist contains social ontological premises that are rather strained and cross-strained. These, in turn, are reflected in the social ontologies of many of those experts which reflect on society. I will dubb one special manifestation of this strain "The Wittgenstein-Problem".

LANGUAGE GAMES AND FORMS OF LIFE. An astonishingly unanimous opinion holds "language game" and "form of life" to be two key terms of Wittgenstein's philosophy. It is common knowledge that Wittgenstein decided, in the most influential version of his linguistically turned thinking (in the second period of his work), to consider language in comparison to a game played according to fixed rules. Systems of communication were more or less related to games in everyday life. (Children anyway learned their mother tongue in games and as games).

The *rules* of these games are not identical with the regularities an observer might be able to detect. Rules naturally cause or support regularities of acting. If somebody acts according to the rules he therefore normally succeeds in doing something *repeatedly* the way others do it. Another famous argument of Wittgenstein contends that a single person never can follow a rule only once and alone [cf. Wittgenstein, 1976, aphorism 199][1]! But beyond the regularities they produce, rules entail a binding character, an aspect of prescription or normativeness. There are at least two ways of socially breaching the rules which are both followed by an outburst of negative sanctions on behalf of significant and unsignificant others. One is intentionally deviating from the norm, the other, unintentionally making mistakes or committing errors. One can follow or use a norm without being able to *mention* it.

All these aspects can also be found in games and their set of rules. Of course the rules of the game that is called "language" are not as fixed and strict as the rules of chess. During their social use in speech-acts the rules of generation, association and adressing of symbols are subject to a constant change, i.e. language is in a constant flux, but a flux that is for Wittgenstein and his followers still steady enough to adhere to the image of rules within certain games — to the metaphor of a game of chess.

"The" language is called "a" game. But "the" and "a" are not as uncomplicated as they appear. A Babel of languages is known to exist. Have they got anything in common? Universalists would affirm that they have and continue to look for a single set of "deep" rules that supersedes the gaudy multiplicity of the different games of language. That is, they would still insist on language (singular) as a game. Universal pragmatics, universal grammar and other approaches looking for *basic* rules of speaking and action or claiming to have found the aprioris of communication or principles of discourse, are all excavating the foundations of a single system called "language". All of these approaches are accustomed to see the linguistic turn as an important caesura in the development of philosophy. The ancients — rather unsuccessfully — sought according to this opinion for the *ousial*, the essences as subtances; the modern protagonists of the Enlightenment — again rather unsuccessfully — dived deeply into the constitution of the constituting subject seeking for universal ideas that generate the world as it appears. But nowadays almost everybody has turned to language — as analysis of language alone seems to promise success[2].

Non-universalists, or regionalists, do not protest against this tremendous twist as such. They only question the excessive rationalism of universalist programs in the face of the seemingly never-ending variety of games of language that only bear some deeper similarity to each other in borderline cases. Language games, they say, have no universal principles in common. Certainly some of them show some similarities ("family resemblance"). But in principle there are no rules to be found which allowed any reasoning about "language-in-general". In some extreme, yet actual cases, regionalism takes individual language games as a kind of hermetic meaning-capsule all of which rank equally in their relation to god[3]. They take them as equally valid and indifferent in their relationship to each other.

Many philosophers support or accompany regionalist tendencies by means of the theory of an uncommensurability of paradigms or by the serious recommendation that only one rule is applicable in science, the rule namely that every rule is applicable: "anything goes" etc. Some sociologists try to comply with regionalist recommendations. They dress, intellectually, in the Haute Couture of French Postmodernity. Into this fabric slogans like: "There are no totalities!" (Lyotard) are woven. This means: There are no structures or processes characterizing society as a whole, there are no fundamental rules which could supersede those separate meaning-capsules. There are ultimately no superseding social structures for regionalized forms of life, as for instance classes once were said to be. (At least this could be taken as the sense of some slogans, the newer waves of this theoretical tendency).

On issues such as these Wittgenstein did not want to make a firm decision. At times Wittgenstein seems to be saying that: "the" language of everyday life is "the" ultimate game, which is the basis of all the singular languages. However, holding this opinion he could not have been referring to anyone of the empirical languages spoken anywhere in every day life. He must have had a system of universal and basic rules in mind — at least if the version of the problem presented here is under consideration. And this version parallels the course taken by Husserl who presupposes a definite *structure* of the "Lebenswelt". (Even though we have to dispose elegantly of possible differences for the moment between Husserl's "Lebenswelt" and Wittgenstein's "Lebensform"!). For Husserl all troubles with pluralism of language-games or life-forms disappear since the life-world has a formal and general structure which "... remains invariant in the life-world throughout all alterations of the relative" [Husserl, 1970, p. 142]. Conversely, there is enough support for the opinion that Wittgenstein went no further in his analysis than the variety of different language games which only in limited cases occasionally overlap. We can stumble over tensions of this kind between universalism and regionalism in many of the writings of linguistically turned sociologists. But even when this problem is set aside, the relation between "language game" and "form of life" remains problematic. The remainder of the present discussion will deal with the Wittgenstein problem in this more general form.

What is a form of life? There is little chance of finding definite answers to this question in Wittgenstein's writings, at least, not answers definite enough to prevent warfare of rival citations and quotations. To replace "form o life" by "human life" would certainly mean replacing a dark expression by an unenlightened one. But if we understand "form of life" — following the *Blue Book* — as a pattern of actions and reactions there will at least be a chance to insist on "form" as an expression for the particular *arrangement* of singular acts ("pattern") and on "life" as social action ("activity"), or a social behaviour ("reactions").

Nevertheless the literature in this field leaves quite a lot of alternatives to be considered: Is "form of life" identical with an arrangement of speech-acts, since speaking is always a way of establishing or cancelling social relations with other persons? Is "form of life" to be equated with systems of observable non-verbal gestures which also have meaning for others? Is "form of life" to imply the original biological equipment of all human kind, as it is this endowment which supports a general and common way ("form") of living? Or is "form of life" to be analyzed from a cultural-historical, let's say: a sociological, point of view [Gier, N., & Gier, F., 1980, p. 241]? The latter would be the case where the idea of "form" approaches, for example, Max Weber's "Stil der Lebensführung" (way or style of conducting one's life). "Life" therefore would continue to comprise actions and reactions in their relation to normative expectations and institutions or processes of society as a whole. "Life" ultimately would coincide with "society". "A bit of everything" is certainly not the wrong answer. But Gier not only wants to prove that the cultural-historical conception of forms of life is the one that is most widely accepted, he also intends to demonstrate that it is in accordance to the assumed intentions of Wittgenstein. The linguistic pirouette is placed on stage with the presupposition that "form of life" may indeed be taken as meaning definite arrangements of actions and interactions in a context of cultural norm, social institutions and societal processes, that "form of life" stands close to the concept of society. This pirouette consists of a restless movement which is choreographed in three twisting movements. The clearest view of these is achieved when we see them as classes of relations between language-game and form of life depicted in Wittgenstein's works.

Relation 1:
    "I shall also call the whole, consisting of language and the actions into which it is woven, the "language-game" [Wittgenstein, 1976, p. 27].

According to this aphorism "language game" (S) represents a unit, a whole, consisting of two main parts. It's first part is called: "the language" — perhaps in the narrower sense of a language game (S^*) seen as a system of vocal utterances according to rules. Its second consists of a pattern of human actions (L = "form of life") with which (S^*) is necessarily "interwoven" to use Wittgenstein's word. The metaphore of being interwoven hints at modes of relations to which the unity of the two parts (S^* + L) are due. Yet this textile image opens two general possibilities, not one.

(1a): Language (S^*) and form of life (L) appear as two parts with *separate* characteristic traits. Nevertheless these two elements are interwoven into a context called "the language game" (S). The picture is one of two components tightly connected by a system of interweaving relations w.

(1b): Language games (S^*) are systems of speech-acts according to rules. Certainly rules are of verbal or symbolic stuff. But rules also imply the essence of actions. Rules not only guide actions, they produce them. In an important borderline case this is going to mean that all acting, every human action that happens, is due to rules as principles of language ($S!$). That is, social action is symbolically constituted action. The relation of interweaving would now point to the thesis that language (S) *is* a system of acts, a medium by which competent speakers not only arrange symbols (S^*) but establish or cancel social relationships (L) — sometimes even relations to things. Greetings, pleas and curses — as we know — do not signify simply vocal or verbal utterances, they establish social relations to the adressee of these modes of address. The picture is now one of an identical medal (S) with two sides (S^* + L). The technique of weaving in this case leads to a tighter product in comparison to case 1a.

Perhaps the *locus classicus* and most determinate version of this latter (1b) view may be found in Peter Winch's *Idea of a Social Science*. Winch's version of this view has considerable impression on sociology and ethnology. According to Winch society (L?) appears as a context of meaningful acts — even though one may sometimes doubt the meaning of what happens in society. Meaningful social actions are rule-oriented actions. Rules therefore constitute definitions of social *meanings* or "ideas" (as a corresponding concept says). Rules guarantee the approximate uniformity of human actions and establish the probability that different people can repeat the same act as time passes. Without rules there wouldn't be any possibility of doing the same as other persons do. But rules always imply the risque of committing an error and breaching a norm.

Because of rules, meaningful social action is *symbolic* action. Action with a sense is symbolic: it goes together with certain other actions "in the sense that it *commits* the agent to behaving in one way rather than another in the future" [Winch, 1960, p. 50]. Being in this way committed to a future course is also a consequence of existing rules: By following a rule here and now, I act in a way which not only excludes a range of other possibilities of action for the moment but also a possible set of future deeds[4].

It is in the context of such premises that the particular twist represented by 1b is defended: According to Winch there is an *internal* relation between rule and action, i. e.: bearing their own empirically and logically irreducible traits. Winch definitely and clearly maintains that social relations between men (L?) and the "ideas" (S*?) embodied in the actions of human beings are "the same thing" only looked at from two different points of view. The difference between a symbolic system called "language" (S or S*?) and a system of interactions, i.e. social relations (L), turns out to be caused by the effect of perspective. Winch insists that it should not be overlooked that "the social relations between men (L?) and ideas which men's actions embody are really the same thing considered from different points of view" [Winch, 1960, p. 121]. The relation of "being interwoven" (w) here, without doubt, is loaded with the whole force of logical implication. Put it this way: To know the content of rules is to definitely know the corresponding "way of life", or system of actions.

While there are indications of discomfort with his own position in Winch's book[5], he provides us with a very clear example of a *social ontology* for whose effects much of the writing in sociology of symbolic interactionists or ethnomethodologists, could bear witness. It is beyond my capabilities to summarise the ontological position in English as pithy as Heidegger's German but maybe the following formula scrapes through as a summary: In case 1b the being of society *equals* the being of meaning (and therefore the being of rules).

Relation 2

"Here the term language-*game* is meant to bring into prominence the fact that the *speaking* of language is part of an activity, or of a form of life" [Wittgenstein, 1976, p. 23].

Has Wittgenstein turned the situation around by 180 degrees now? Language games (S*) (understood as speaking a language) are now understood as *part* of an activity and this means: of a form of life (*L*). This in turn offers again two possibilities.

(2a): First of all the relation of being interwoven could now be understood in accordance to a relation between part and whole. In this case the form of life (*L*) would appear as a sort of totality that contains the language games as subsystems. Some of Husserl's arguments concerning the overarching character of the "Lebenswelt" seems to lead in this exact direction.

(2b): On the other hand the term "part" in Wittgenstein's sentence could hint at something like the relation of intersection between S (*S*) and *L*. This means: *S* (*S*) and L *are not* identical but there exists an intersection of common traits. In this case the thesis of an *internal* relation between *S* (*S*) and *L* can no longer be supported.

Relation 3
"Giving grounds, however, justifying the evidence, comes to an end; – but the end is not certain propositions' striking us immediately as true, i.e. it is not a kind of *seeing* on our part; it is our *acting*, which lies at the bottom of the language-game" [Wittgenstein, 1969, nr. 204].

This aphorism is interesting for two reasons. First of all a new twist takes place. As in the case of the part-whole-relationship (2a) the language-game is ranked behind – or even below – the "actions", respectively the "behaviour". Language-game (*S*) is ranked below action and therefore form of life (*L*) as a system of actions precedes it.

But this account of the relation is further aggregated in Wittgenstein's own language-game by the concept of foundation ("Grund"). It is now assumed that the roots of the language-game can be found in action or in prelinguistic behaviour. In other words: The form of life (*L*) is the basis for the language-game (*S*). Especially in *On Certainty* Wittgenstein's theory of language-games is diving for the deepest foundations. At this point we need to be free of any susceptibility to vertigo, for another twist could be ahead: "You must bear in mind that the language-game (*S* or *S*) is, so to say, sometimes unpredictable. I mean, it is not based on grounds. It is not reasonable or unreasonable. It is there – life" [Wittgenstain, 1969, nr. 203]. Now, suddenly it is possible that "the (*S*) or "a" (*S*) language-game has reached the point where all questions concerning: where from?, where to?, what's behind?, is it reasonable or unreasonable?, have to stop. In the context of relation 3 I am able to imagine S only in so far as: a) it characterizes the unity of the two different components S' and L; b) the relation uniting S^* and L points at the same time to one of the two components as a principle.

That is to say: There is no path beyond S as *unity* of two nonidentical elements: The links in the chain of reasons come to an end, at the boundary of the game. In this manner S would indeed be "unpredictable". Nevertheless one of the two elements may keep be the more basic. Or — Hegel would put it despite Wittgenstein's frown — one is seen as "essence", namely L, the other, namely S^* as "appearance". L means the basis and even the language-game (S) stands there like life — life remaining the basis that is meant.

I am not seeking to make any outrageous connections with Hegel's logic of essence and appearance which might be open at this point. I'm interested rather in sociological and social ontological positions which express *this* twist: Marx and Mead have something to say on this. It should be remembered here that they both agreed with Wittgentein's famous argument concering the impossibility of private languages. Marx briefly touches on it at a point in his *Grundrisse* where the structure of relation 3 also quite clearly appears too.

Mead supports relation 3 with a short and striking statement: "... so far from being a precondition of the social act, the social act is a precondition of it" [Mead, 1934, p. 18]. The other statement: "You cannot convey a language as a pure abstraction; you inevitably in some degree also convey the life that lies behind it" [Mead, 1934, p. 283] stems from Mead, not from Wittgenstein.

Even though those who would use the term "social ontology" have to reckon with reservations on the side of linguistically turned minds, it is clear that Wittgenstein's theory of language-games is deeply involved in matters of social ontology. But it is not easy to decide which ontology he ultimately meant. As far as relation 3 is concerned there is a hidden path leading to Marx and Mead. At the end of this path "life-form" stands as ontological basis of the language-game or the language-games. In Marx' work "life-form" means historical existing types of society and the way Marx conceptualizes these societies may not be acceptable for a true Wittgensteinian. Nevertheless "society" or "social systems" as they really exist are indicated as possible proxies for the founding component in relation 3. Wittgenstein once defined the basic "life-form" as "pre-linguistic behaviour". This could immediately trigger a quarrel over the question of whether Wittgenstein had the same thing in mind as Mead: an observable behaviour of "organisms" (Mead) which — through gestures — entices a more or less significant other into reactions.

Naturally there are different theories of "society" which could provide a concept of "life-form". The point made here is simply this: Relation 3 places "form of life", in the version in which its content equals society onto centre stage. Thus, "form of life" is the point where all questioning has to end.

The reader may well question the value of dealing with these three relations in a manner that recalls Beckmesser's procedure in the "Meistersinger of Nuremberg". Discomfort may further be increased if their seems to be an iclination on the part of the author to squeeze theoretically significant meaning out of each trivial sentence, following the Oxford or Cambridge style. My response to this: In my view all twists and turns connected with the Wittgenstein-Problem are implanted to an astonishingly high degree in the paradigms of recent sociological theories. This may be due to the constant reappearance of all the three figures in the context of one and the same sociological theory, or it may be due to the fact that one approach is internally fractured into fragments consisting of the various movements of the pirouette.

EXAMPLES FROM FOUCAULT AND ETHNOMETHODOLOGY. I will describe *two examples of pirouettes* in social philosophy and sociology in support of the present argument.

All the many words uttered in society cannot — according to Nietzsche [1974] — represent the single, totally individualized "original appearance" they are based on. Concepts rather have to fit "a lot of different cases". Thus they unavoidably reduce the individuality of details — says Nietzsche. It is not their relation to things but the condition of human life that gives them significance. Those brilliant animals on our planet who have discovered reasoning may consider themselves capable of seeing into the essence of things as they are, but for Nietzsche, ultimately, all paths lead to seeing the intellect primarily as a means for individual subsistence. As such, it must also be seen as a means of deception and camouflage, aimed at the survival of man-kind. However, there is also a need for a peace treaty, survival demands that we get rid of the worst effects of a belligerent state of nature. A core of common regulation of language is necessary for such a peace treaty. That means, according to Nietzsche that humans must discover a mutually valid and binding denotation of things. With such systems of language an antagonism is set up between truth and life. This is certainly not the same thing as saying a partly pacified humanity could from thereon be able to see into things as they are and hence be able to use language as an "adequate expression of all realities". The distinction, or — to put it in a more modern way — the codification of true and false is, according to Nietzsche, rather based in the individual's will for survival. And these individuals are forced by need and boredom to exist socially and in packs, he says. To offend against language-games and their "metaphors in use" (Nietzsche: "usuelle Metaphore") once they have been accepted only heightens the human life-risque. The refusal to join usual language-games, i.e. the decision to codify true und false in a way that departs from usage, can only be defended at the prize of "offending the life-preserving consequences of truth". Such linguistic spoil-sports have to be prepared to suffer for it, and to take into account discourse-police who maintain their sentries in academia.

Those constraints and forces that throw themselves into the path of any lingusitic spoil-sport often appear, in Nietzsche's text, as the retributive actions of significant others. Simultaneously, he asserts the domination of language-games as systems, or structures, over living people. In this sense, the constraint does not stem from persons but from the abstractions themselves: the example of the habitual liar, whom nobody trusts, and who is shunned by everyone, manifests the utility of thruth-codes for human life. We are not able to act as rational beings without placing our doings "under the authority of abstraction". According to Nietzsche's linguistic theory, stimuli of our nervous system are transformed into images, images are moulded into sounds, yet these sounds freeze into "conceptual fetters". In his opinion these conceptual fetters can never represent the "exclusive, totally individualized original experience". The "power of abstraction" is seen by Nietzsche as the anonymous force rules execute over the unrepresentable singularity of an impression as well as over thinking and acting of singular persons. This power or authority is contradictory in itself. On the one hand, it points back to the above-mentioned societal basis of speaking the truth. This means for Nietzsche that it points back to the obligation "to be truthful" in order to enjoy a comfortable life. On the other hand, Nietzsche also describes this obligation as the readiness of individuals to keep to the "metaphors in use". But as time passes these metaphors become worn like old coins. To keep to them means: to lie in a style obligatory for everyone and — in order to avoid troubles — to put up concepts over the singularity of impressions. In the last analysis, the conceptual fetters stand opposite

> the perceptual world of first impressions assumes the appearance of being the more fixed, general, known, human world [Nietzsche, 1974, p. 181].

The conclusion to be drawn is this: An anonymous will for power Nietzsche sometimes calls "the regulating and imperative", is rooted in the strict conceptual order itself and not in the motivation of any planning subject or institution. The will for power ultimately appears as an attribute of officially used language-games.

In arts as well as in myth, Nietzsche thinks, the fundamental human drive to produce metaphors — a drive never to be tamed or suppressed — finds its way out of the cages of conceptual order.

> This impulse constantly confuses the rubrices and cells of the ideas, by putting up new figures of speech, metaphors, metonymies; it consistently shows its passionate longing for shaping the existing world of waking now as motley, irregular, inconsequentially incoherent, attractive, and externally new as the world of dreams is [Nietzsche, 1974, p. 188].

Nietzsches theory of language may appear somewhat antiquated but Michele Foucault explicitly accepts and defends two implications of his argument while parting company with another.

First, Foucault does not describe the "production of truth", as an enterprise calculated to gain insight into existing things, but as a socially significant and socially consequential praxis of "coding". Exercises in the distinction between true and false always imply the power-guided separation of sheep from goats and therefore no glimpse of the things as they are at all. In agreement with Nietzsche, for Foucault "truth" is no realm of true objects to be discovered. Here is only a set of rules in order to separate "truth" from "falsity" on the basis of power [Foucault, 1981, p. II]. It may be that arguments do have a binding "force". But according to Foucault this "force" has not much to do with the possibility that the *logic* of an argument may be *empirically* followed by people being impressed or moved. The "force" consists in the power of "the discourse" over the individuals. *Second*, the anonymous system of rules that guides "coding" and produces its power-effects through "production of truth" has not much to do with personal or institutional centres of social power. This power of discourse is not one like that of a subject or a group or class acting. The force of compulsion and its endowment with power belong to discourse itself. They are implanted in discourse or represent its effects. *Power*, a central term in Nietzsche's and Foucault's work as well, is in their case wrongly rendered in the singular. It is more correct to see discourse as a field of force, as a manifold of power-relations, which organize a region [Foucault, 1972, p. 215-237]. Hence truth too can have power. On the one hand truth is produced by various constraints, on the other hand — as practice of bifocal codification — generates regulated effects of power. These can be said to appear where no coercive discourse-police, but "discourse itself", stringently defines where we have to do something or to leave it undone. Finally the production of truth builds up into a game with an immense machinery of exclusion which imposes on the subject a certain position, a certain view and a definite function [Foucault, 1972].

All this shows that language games dominate living subjects in a manner corresponding to Nietzsche's vision of imagination overwhelmed by abstraction and concept. For, Foucault says, we have to see discourse "as a violence that we do to things, or, at all events, as a practice we impose on them" [Foucault, 1972, p. 229]. However, in a third respect, Foucault wanders from Nietzsche's path. There will be not much doubt that Nietzsche relates language to society or life as its fundamental principle. The sentences he uses in this context could — in a modified form of course — be found in some of Mead's works. For example the statement:

> "I may proceed further to the conjecture that *consciousness generally has only been developed under the pressure of the necessity of communication* – that from the first it has been necessary and useful only between man and man (especially) between those commanding and obeying) and has only developed in proportion to its utility ... In short, the development of speech and the development of consciousness (not of reason but of reason becoming self-consciousness) go hand in hand" [Nietzsche, 1910, book V, aphorism no. 354, p. 298].

Maybe it is the philosophy of survival as an undercurrent in Nietzsche that provided the reasons for Foucault's deviation from this sort of alignements. But the effect of parting company with Nietzsche is manifested by Foucault continually passing through all three postions of the pirouette — with "society" and "discourse" as alternate pivots: Foucault's own impression of "discourse" as a concept endowed with a meaning "rather at sea" is not going to provoke vehement protests. His famous definition of "discourse" as a class of statements belonging to the same discursive formation [Foucault, 1972, p. 17] has a touch of tautology. But this impression may be weakened. He only wants to point out statements and groups of statements as the matter the discourse is built on. Ultimately he is interested in the principles which underly the formation of statements, and these principles count "rules" among their decisive parts — just as Wittgenstein says. Yet with Foucault statements appear as "enonces", as power-loaded speech-events. The activity of producing statements and utterances principally depends on the game of the forces and powers that govern the appearance of any disclosure (enoncé). Another decisive part of the definition is the gap which inevitably appears between the rule for a statement and the actual speech-act. Discourses thus appear as supraindividual unities of power and acts of disclosure whose basic elements are to be seen as statement-*events*. In this context is posed one central question of Foucault's theory of discourse. How is it possible that a discourse assumes "an exact specificity in its occurence and assumes a place that no other could occupy" [Foucault, 1972, p. 28]? Our question in the same context reads: How are *discourse* and *society* related to each other? Foucault definitely chooses all three movements of the Pirouette:

MOVEMENT R1. Foucault says discursive practice incorporated a whole, i.e. a unity of anonymous rules that can be identified in space and time. These rules are said to regulate the appearance, the power-effects and conditions of operation of the "enunciative function" during an epoch [Foucault, 1972, p. 117]. Thus the notion of "discourse" approaches closely the concept of "language-game" (S^*). "Discourse" thus seems to aim at the gaudy historical variety of discourses (S^*) in reality. Nevertheless Foucault continually uses the term "the discourse" ($S?$) as a general notion overarching the singular discourses (S^*). I believe that the drawing of parallels between Wittgenstein's "Lebensform" and Foucault's "society" is sound (L). The first figure of the pirouette (in its sharpest version 1b) surfaces therefore in all those cases where Foucault considers a model of imposed subjectivity as respectable. I.e.: Whatever the individuals are and what they do according to this model ultimately has to be considered as an effect of "the discourse" (s) or of the "discourses" (S^*). In this manner the "reduction of men on the structures surrounding them" [Foucault, 1978, p. 16] indeed becomes the central topic. But these "structures" appear as *systems* of discourse-forming rules to which the rupture between rule and actual speech-act belongs itself as event of "la différance". Structures ultimately posit (condition) the disclosures and actions of the individuals. What in Winch's reflections appears as a relation of *Implication* between Language-Games (S^*) *and* social form of life (L) would now appear in Foucault's work (of this period) as an effect of powerful discourses which form the disclosures and actions of human-kind.

Relation 2:
"I am supposing that in every society the production of discourse is at once controlled, selected, organized an redistributed according to a certain number of procedures, whose role is to avert its powers and its dangers, to cope with chance events, to evade its ponderous, awesome materiality" [Foucault, 1972, p. 216].

This statement is open to both variations of the second turn of the priouette (= 2a; b). The discourse, or the discourses, could be part of the society which represents an all embracing unity. A certain inter-section between discourse and society could be meant as well. Each society has its regiment of truth, its general politics of truth, i.e. types of discourse, which it considers as true and implants as truth [Foucault, 1981]. In this case "society" (L) contains the several types of discourse (S^*) in itself. In both cases, however, at least a minimal stock of criteria of difference between "society" (L) and "discourse" ($S; S^*$) is claimed.

Relation 3:
From time to time Foucault supposes a relationship of foundation between "discourse" and "society". Again at least two possibilities come up: On the one hand "discourse" (S^*) and "society" (L) appear as two different components *of* the unity of "the discourse" ($S!$) — respectively of "the society" ($S + L$). On the other hand S^* and L might be seen as two separated parts which stand in a certain relationship to each other. The decisive point here is: In both cases a relation of foundation, a relation to a ground or basis is supposed. "But this will to truth, like other systems of exclusion, relies on instrumental support: it is both reinforced and accompanied by whole state practices such as pedagogy — naturally — the book

system, publishing, libraries, such as the learned societies in the past, and laboratories today" [Foucault, 1972, p. 219].

Now at last the great game of truth-production has a societal basis! Or is there any better reading of statements of the type: The production of "the discourse" would be controlled in society (as the substratum)?

It may be that Foucault intentionally did not want to rest with one of these three "epistemes" (R1 − 3), may be he wanted to rotate them without ever stopping? I dare to surmise that those three twists cause a lot of the trouble which occurs when one attempts to make use of Foucault's theory of discourse. The arbitrariness especially of the presently fashionable use of his concept of discourse in sociology might lead back to the movements of Wittgenstein's Pirouette.

2) The hermit Serapion's brethren[6], the ethnomethodologists in sociology provide a second influential example. This example, however, involves the breaking down of an otherwise unified paradigm in the sciences of society. The fragments produced depend on the foothold chosen for the pirouette as might be sufficiently demonstrated by looking at only two of the central concepts of modern ethno-methodology: "accounting practices" and "mundane thinking".

"Accounting practices" may be considered as activities governed by rules, as methods, which "we" as actors in everyday life are accustomed to use. But by using them we not only produce these "Lebenswelten" we make them apprehensible to other actors. I.e.: ethnomethodological studies see everyday activities as methods whereby the members of a society produce these same activities, making them communicable for all practical purposes and present them as "rational" as well. These rules steer the organization of common practices in everyday life and at the same time allow duplication of them. In this sense they become "accountable". The decisive medium for this queer accomplishment is handed to us in everyday language. Should the little green fellows again start from Mars they could register with astonishment that the aborigines are engaged in a neverending process of describing and explaining what they have done in the past [Wieder & Zimmermann, 1970, p. 105]. "Accounting" seems to carry a triple meaning: 1) The procedures in everyday life (S^*) "represent" the life-forms (L), i.e. they *are* or *produce* them! 2) They represent the life-worlds (L) through utterances in everyday language (S^*) *for* others, they make them accountable *for them*! 3) Those procedures make the life-worlds apprehensible as "meaningful", at least as usual ways of acting, or even as "rational" proceeding.

In contradiction to Parsons the rules of procedure are not seen so much as normative principles the individual would have to introject in order to guarantee a comparable and "well-situated" activity. Using these procedures our life-worlds are continually reproduced *anew* and even in a *new way*. In this manner society is not seen as an integrated *being*, but as an collective *product*. This result is obvious enough, yet it opens up very different paths for social ontology: e.g. the "product" could at core show all the attributes of a general *process*. And this process could present a "logic", an order of its elements, which indeed indicates a regularity but might not be reduced without rest to the existence of rules and methods in everyday life. Naturally an opinion like this has to be contradicted vehemently by real anchorites, i.e. ethnomethodologists who look for *relation 1* as their foothold. In clearer and more definite cases they would suggest that the "activities whereby members produce and manage settings of organized everyday affairs are identical (!) with member's procedures for making those settings »accountable«" [Garfinkel, 1967, p. 1]. In other words: Everyday language is not only "reflexive": when making use of everyday language, through speaking it, a form of life is *posited* at the same time. "The setting gives meaning to talk and behaviour within it, while at the same time it exists in and through that very talk and behaviour" [Leiter, 1980, p. 139]. A verbal reading of all the many statements pointing exactly in this direction in the texts of Serapion's brethren leads to an unavoidable impression: Representations (accountings) (S^*) in everyday life are *identical* with the production of forms of life in everyday world. The foothold is R1: $S^* = L$!

Reflections on "mundane thinking" can deliver further clarification of the influence of the three foundamental movements a Wittgenstein-ian Pirouette commands. "Mundane thinking" is introduced by Pollner and other authors as a system of evidential beliefs our thinking and act-ing in everyday life is oriented to: A mundane thinker socialized in the normal way does not only take an objective world as given, he takes it for a world others can approach too [Pollner, 1970, p. 296].

The structure of our common language does not clash with the first movement of the pirouette. It is comparable with a realism or material-ism which is sometimes scolded as "naive" by advanced philosophers. This system of beliefs is not only characterized by the obstinate thought that there are things exterior to our language-games but also that there are social facts being more than the mere product of speaking, using meanings of following the rules of any forceful language-game or discourse. Isn't it reasonable, even for a philosopher, in his own everyday life, to surmise that the motorized thin-box which had just appeared on his side of the highway is of a more solid stuff than meaning, sense and linguistic rules [Mehan & Wood, 1975, p. 102]? Certainly realism in normal life is not without problems. The characteristic tendency to neglect the self has often been criticized [cf. Hegel, 1979, vol. 4, p. 111]. Yet where sociological matters are concerned a true anchorite has no room for compromise with mundane realism. He has to insist that the whole social reality the naive mundane thinkers (we all are in a way) dare to deal with depends on rules or methods which are unseparably interwoven with the common language-games (S^*) mundane thinkers are accostumed to use. Indeed some ethnomethodologists tend to underline strong statements: "The sense of social structure is the perception" and the assumption (!) that the social world is a "natural order" — "Social order appears *on* gestures and *in* utterances"[7]. How to deal therefore with mundane thinking?

THESIS: Events and processes existing "factually" or "objectively" do belong to society, i.e.: They exist "exterior" to our language-games ($S + S^*$). They may be influenced by them, but they are not posited by them or reducible to them. In this case it would be consequential to say: "A natural order is objective; it is "out there" for people to see and interact with" [Leiter, 1980, p. 69]. Without doubt society is a PRODUCT of human action. But this product is characterized (e.g.) by synthesizing *processes*, which can contain sense, meaning and rules, but — *nevertheless* — are not *constituted* by linguistic principles.

ANTITHESIS: Individuals in everyday life *think* that there exists a factual social order of structures and processes. The laymen *imagine* social reality as a world *sui generis*. All an ethnomethodologist undertakes is to discover the methods with which mundane thinkers accomplish the picture of a "factually" existing social world. Society *is* — whether seen as a product or a process — a realm produced by meanings, symbols, rules ...: The following sentence might exemplify this position: every reality depends upon:

1) ceaseless relexive use of
2) a body of knowledge in
3) interaction [Mehan & Wood, 1975].

But another possible reading of the same statement could trace the tension between thesis and antithesis in it: "Knowledge" (S^*) is used *in* "interactions" (L). Logically the two possibilities mentioned stand open: S^* and L might be identical, or they might show some traits *differentiating* them. The latter reading would lead to position 2 of the pirouette. For this example the location of the foothold is not easy to decide.

One faction among the ethnomethodologists seems to support *position 3*: This has something to do with the influence of Edmund Husserl on the phenomenological school of A. Schütz and its reconstruction by ethnomethodology. It goes without saying that Husserl emphasized that the everyday "Lebenswelt" (L) has a principle, or a foundation. The "Lebenswelt" is continually inserted in his analyses as "Untergrund" (founding layer). This kind of relationship, of language-games (S^*) to life-world as their "ground", is often chosen by him to illustrate the standing of scientific games of language in regard to everyday forms of life. It is one of Husserl's central theses that even the semantical content of scientific language-games *necessarily* implies semantic components virulent in everyday life[8].

Naturally all problems are connected to the attributes of the "Lebenswelt" now. Unfortunately, I must refrain at this point from reflections on the broader relation (and difference) between Husserl's analysis of life-worlds and Wittgenstein's theory of life-forms. However, there are two social ontological themes — rather rich in consequences — that should not be forgotten. In a way they recall the thesis and antithesis in ethnomethodological dealings with the problem of mundane thinking mentioned above. I think there will be little protest when I say that Husserl preferred in his *Crisis* a *univeralistic* perspective on the *founding* Lebenswelt. "Lebenswelt" is not identical for him with the free play of unsurveyable provinces of meaning. The "Lebenswelt" has a "structure", he says [Husserl, 1979, p. 142]. Thus "Lebenswelt" is still taken as a totality. And within this totality "structures" may be discovered. In my opinion, Husserl elevates these structures to the rank of universal "constituting principles" or "aprioris", as he calls them. Some of Serapion's brethren seem to have taken up these universalistic, reality-producing structures and transformed them into speech-act-generating fundamental *rules*. In this sense they enter discussions, e.g. in the form of Cicourel's "basic rules". Now, this turning to basic rules once again opens two possibilities: first the pivot of movement 1 of the pirouette could be implied. In this case basic rules as linguistic factors would ultimately guide the production of life-forms. (Movement 2 is also imaginable). But only when the "Struktur der Lebenswelt" is seen as a ground which does not depend only on linguistic traits, are connecting paths to Marx' and Mead's visions of a "material" — Mead would say: "behavioural" — ground opened up. At one place Husserl says it would not be reasonable to substitute "being" as science understands it for the being of the life world [Husserl, 1979, p. 124]. But this does not exclude the possibility of thinking the "being" of society (L) to be something more than a being through meaning, sense and language. This causes no conflict with Husserl's insight that social facts (the structures of the Lebenswelt) should not be treated like material substances with primary and secondary qualities. But the true hermits in ethnomethodology have to reject *all* strategies to elevate some traits of the life-world to the status of universal "structure" founding language-games (S*), even though the topic: "social structure in ethno-methodology" remains a burning issue: "It is my contention that every reality is equally real" [Mehan & Wood, 1975, p. 31]. Now indeed the pirouette has not only turned back to its beginning, it has brought us face to face with the problem of the pluralism or relativism of forms of life, itself connected, in various ways in each of the movements of the pirouette, with pluralism or relativism of language games discussed earlier in connection with the linguistic "regionalists".

WAYS OF DEALING WITH THE WITTGENSTEIN-PROBLEM. In my opinion it is not possible to solve the Wittgenstein-Problem by mere philological techniques. There is always a good chance of finding a quotation somewhere in Wittgenstein's work which favours one or the other interpretation. One need not be depressed about this. Great Philosophy often proposes different perspectives upon one issue, perspectives which can easily lend themselves to expansions in one direction or another. Certainly the most elegant solution of the Wittgenstein-Problem is to remove it by fiat. For example, one needs only to expand the concept of meaning in sociology so much that it ultimately coincides with the process of system-building or forming a society itself. Then — *per definitionem* — there will be no alternative to the first movement of the pirouette! Generally all the three movements play their explicit or implicit role in sociological approaches influenced by the linguistic turn. Quite often they appear in form of endlessly twisting pirouettes in relation to which the declaration of a firm stand has no firm standing. The rather facile use made of the concept of "discourse" in fashionable sociology might hint at that.

My way of dealing with the problem? I do not think the problem is a linguistic one which could be solved by squeezing the last drop of meaning out of sociologist's sentences. The role the ubiquitous "Lebensform" points back to the tacit differences in social ontologies characteristic of the different approaches discussed. It's interesting that they seem to be imported *as* social ontologies from philosophy into sociology. Afterwards they appear as mere traces in social scientific approaches for which a linquistic turn is evident. Without doubt the consequences of such deeply rooted presuppositions filter into the most remote corners of sociological research. But even the most artful performance of the pirouette has — while passing through all three postions — a certain limit. Not all ontological positions can be displayed in one act. To the contrary, one has to reckon with solid contradictions between them! To point out these contradictions and to work on them could be first step in the direction of dealing explicitly with social ontological premises no sociologist — in my opinion — can neglect.

At this stage in my argument I have only a limited opportunity to sketch such steps by assembling a few sentences and counter sentences which support or reject *position 3*. For myself, it is the postion I dare to prefer. (The reasons for this predilection have a lot to do with my opinion — which I can't expand on here — that this would be the best way of dealing with a central topic of sociological reasoning which leads back to Hegel, Marx and Mead. A topic which might be condensed into the question: How is it possible to refer the structure of the general process of economic reproduction of a totality back to concrete systems of interaction which — while indeed being deeply influenced or even constituted by symbols, rules and language — are a decisive factor in the development of a "self"?)

THESIS 1.: Without doubt there are actions and systems of actions (L) in social reality which are dependent on rules of language (S or S^*). The so called "Thomas-Theorem" refers to this: Some social phenomena are "real" when they are "defined" as real" by the actors themselves. There are other versions of this thought more elaborated linguistically. Searle's works provide infuential examples [Searle, 1969], for Searle once insisted that "constitutive" rules for speech-acts could be discerned from "regulative rules". Constitutive rule constitute the *existence* of the behaviour that they simultaneously regulate. There could not be a game of soccer as a societal praxis without the set of rules codified by the FIFA. "Game of chess" is nothing without it's rules.

In contrast to this, "regulative rules" guide forms of behaviour existing independently from them. In this little word "independently" all the problems of social ontology are now concealed. At least, three possible readings could be chosen: a) There are traits of actions and systems of actions which can be described or explained independently from *those* constitutive rules that guide them. Nevertheless these traits could be the expression of *other* constitutive or regulative rules. b) There are characteristics of actions which can be reduced neither to regulative, nor to constitutive rules. Nevertheless it would be possible to understand these characteristics as *regularities* discovered from the observer's point of view, whereas their basis continues to rest in the orientation the actors themselves have to rules and meanings. c) The strongest thesis runs: Traits and "structures" which may not be reduced without rest to the rules of language or to ideas belong to the field of human action. They belong to a form of life. In my opinion thesis 1 concerning relation 3 above coincides with this latter version. But movement 1 of the pirouette is definitely rejected thereby!

ANTITHESIS: "If social relations between men exist only in and through their ideas, then, since the relations between ideas are internal relations, social relations must be a species of internal relation too" [Winch, 1960, p. 125]. — "All forms of communal human life are forms of community of language" [Gadamer, 1986, p. 422].

THESIS 2: "Game of language" (S or S^*) and "form of life" (L) are two key concepts of the Wittgenstein-Problem. Their content might be enclosed in broader or narrower boundaries. Sometimes it is expanded so far that language-game and form of life are not only identified (position 1) but also carry the same meaning as the concepts "society" or "socialization" (*Vergesellschaftung* as $S + L$). The circumstances are further complicated as a more general concept of language-game (S as "the language") and/or a more general concept of life-form (L as "the" world of life) are confronted with the special language games (S^*) and/or special life-forms (L^* as provinces of meaning and action). This opens the possibility of manifold combinations. But which of them may be chosen and expanded — sometimes even Wittgensteinians presuppose a relation of *foundation* between one component ("the ground") and the others. I hold a version of position 3 which implies the following view of a founding relation between L and S to be the most reasonable.

The form of life (L) is seen as "society". It is seen therefore as a historical totality or historical type of society such as capitalism. "The language" (S) or "the languages" (S^*) are *not* identical with the life-form. Moreover there is a founding relation between them to be respected (L-f → S). There is no contradiction to be feared when saying: In the frame of this relation and under its presuppositions there are "constitutive" functions of meaning and rules! Mead repeatedly falls back on the general content of the relation in order to circumscribe the direction of his research-programmes: "We want to approach language not from the stand-point of inner meanings to be expressed, but in this larger context of co-operation in the group ($=L$) taking place by means of signals and gestures" [Mead, 1934, p. 6].

As we have already seen, there are parallels with this opinion in Wittgenstein's thinking:

> "Giving grounds, however, justifying the evidence, comes to an end; – but the end is not certain propositions striking us immediately as true, i.e. it is not a kind seeing on our part; it is our acting, which lies at the bottom (!) of the language-game" [Wittgenstein, 1969, Nr. 204].

Husserl, in exact correspondence with this, deplores how rarely science normally inquires "... into the way in which the life-world constantly functions as sub-soil, into how its manifold prelogical validities act as grounds for the logical ones, for theoretical truths" [Husserl, 1970, p. 124]. Together with thesis 1 (see above) it has now to be asserted that these "prelogical evidences" of Husserl — insofar as they point to the structures of the life-world — are *not* reducible to rules, meaning and ideas! The second thesis on the present level of generality therefore demands a relation of foundation between L and S (L-$f \rightarrow S$) and claims at the same time "factual" or "material" structures of the form of life.

ANTITHESIS: Wittgenstein's views are mobilized for counter-positions as well: "There are many various games of language — that is the heterogenity of the elements" [Lyotard, 1982, p. 8]. This may be a statement of fact. But in its most decisive version the heterogeneity of the language-games is seen as their pluralism to be looked at from a standpoint of ethnomethodological indifference. I.e.: The singular language-games remain with respect to any criteria as equal and indifferent in comparison. Any attempt to make out "the form of life" or "forms of life" as a *basis* or *ground* are nothing more than the impulses of an ailing, old Eropean narration which tries to seduce us into searching for fictitious worlds behind things.

THESIS 3: There are different approaches in sociology which work with the thesis presented: in its most general terms that the life-world is not identical with the game(s) of language but is to be seen as their "ground". At a more detailed level of analysis of this there is a tendency for the various approaches to scatter in different directions. Nevertheless, given the presuppositions of thesis 1 and 2, one road is barred: When descending to the deep ground, structures have to be found lying deeper even than basic rules or constitutive meanings. Otherwise one would be turning back to the first mevement of the pirouette! It is claimed, therefore, that "objective structures" of the lifeworld is a reasonable concept.

I cannot analyze this concept within the present limits of space. Once again, only hints of the directions in which steps could be taken are possible. One sequence of steps — extremely unfashionable at present — leads back to Mead, Marx and Hegel.

Mead normally is called as witness in the case of an interactionism understood in terms of social psychology. His "generalized other" for example is often seen as a term for the groups of significant others with which an individual maintains his interaction which are normally guided by symbols. The interior expressions of the exterior impressions of these significant others on the individual make up the "Me". On the other hand the "I" is seen as competence of the individual to engage practically in the order of the "Me" and (therewith) in the system of established interactions with significant others. But the generalized other is not unavoidably the ultimate point of reference of Mead's statements. Mead explicitly underlines the fact that his theory attempted "to explain the conduct of the individual in terms of the organized conduct of the social group, rather than to account for the organized conduct of the social group in terms of the conduct of the separate individuals belonging to it" [Mead, 1934, p. 7]. But this group does not always coincides with the set of significant others directly related to an individual. On the contrary, Mead wants to depart from society as a whole that is "prior to the part (the individual)". Therefore the part "is explained in terms of the whole, not the whole in terms of the part or parts" [Mead, 1934, p. 7]. And this starting-point is explicitly labelled an "on-going process" [Mead, 1934, p. 7]. Consequently Mead explicates this "whole" as an *organized* totality. But in this case "organization" points to a *process* (or processes) of socialization (*Vergesellschaftung*), to processes which grant the unity of a society, and which are its "ground".

Social process indeed is the term providing the most reasonable point of departure for an explication of what could be meant by "objective" structure(s) of the life-world.

Naturally social processes are *products* of human praxis. Evidently symbolically mediated interactions and the orientations to rules and meaning on the part of the actors do play a decisive role in social processes as well. One need only read the importance Marx attributes to the recognition of certain "fictions of law" for the partial phases W-G and G-W in the circulation of capital. But it is not a necessary consequence of this to claim that the system of relations (in time) between things, deeds, events, phases ... which characterizes a social process could be ultimately reduced to meaning, rule and language. Social processes (at the level of the social system as a whole) show peculiar traits that characterize them *as* processes. One of the most important attributes of this kind connects with the general "logic" of a social process. Processes which guarantee the existence of totalities during a period of history, processes of *reproduction*, necessarily show the quality of *self-reference*. But "Self-reference" is also the central *material* theme of Hegel's *Logic*. And even in Hegel's texts there is not one possibility alone opened up, namely to treat "self-reference" by an analogy to a general subject dealing with itself (Spirit; Self-Consciousness). There are plenty of opportunities to relate his central topic to the modern discussion about "auto-poiesis"[9] which is explicitly seeking the structure and logic of processes on the level of the system itself. Naturally it is Marx much more than Mead who puts the *economic process of reproduction* into the center of his statements about the basis of the Lebenswelt (society). The economic process of reproduction is seen by Marx as the event which "organizes" a totality, forms it — in spite of all it's interior crisis — into a unity. In this sense Marx sometimes calls the "Wertgesetz" (law of value) the "inneres Band" (interior bond) of Capitalism. This interior bond is conceptualized as much more than a regularity which a student discovers "from the point of view of the observer" with respect to the actions of agents oriented to meanings.

Thesis 3 therefore claims: The "objective structures of the world of life" (society) consist in principles constituting the form and/or the ongoing of the economic process of reproduction which "characterizeses an epoch" (Marx).

But let us not confine ourselves to the horrors of Neo-Marxism: normal economic textbook — impregnated with other political options — do not operate on any other social ontological presuppositions. Evidently trade cycles are not independent of deeds and doings of individuals, naturally the famous "social climate" plays its part in the ups and downs of economic growth, and a depression is certainly not simply due to a disturbance of a language-game.

ANTITHESIS: "It is my contention that every reality is equally real" — Ethnomethodologists "claim that member's procedures with the events or structure they describe. Symbolic activities both describe and organize the thing described" — "Social order exists. But its existence always appears within interaction. The problem of social order is thus not a problem of constructing structures about the external world. It is instead a problem of understanding how the sense of a corresponding external world is accomplished by every interactional activity" [Mehan & Wood, 1975, pp. 31, 153 and 184].

THESIS 4: I have claimed, certainly not shown, that it would be possible to mark "objective" structures of the life-world as traits of "ground". This "ground" is seen as a process. To retreat from this postion might provoke a significant misunderstanding: All too easily it could give rise to the impression the "structures" were the "bottom" of all other societal and/or linguistic phenomena insofar as the latter "stem from them" or even are (causally) "conditioned" by them. It is interesting that this dubious view may be found even in authors defending position 1 of the pirouette! Naturally they turn things in the opposite direction: They chose "the language" or "the discourse" as the ultimate point of reference and either identify it with form of life or take language (discourse) with its rules for a constraining of life. A certain theory of the individual and individualization is often combined with this. The singular self appears as an effect — sometimes a forced effect — of rules of discourse. It appears as the place where social threads are connected [Simmel, 1958, p. 2].

This perspective on the problem of the individual is to be found in many sociological approaches especially those influenced by French theory of discourse. Given the presupposition of the language-game (S) or the language-games (S^*) as ultimate points of reference of the theory, the singular person appears as the shadowy figure of a nexus for system of rules in a self-speaking discourse. It appears, so to speak, as a relais for the switching of communications: "Young or old, man or woman, rich or poor, it (the self — J. R.) always is positioned on »knots« in the net of communication, may these knots be as important as it were. It would be better to say: it is posed on posts passed by informations of a different nature" [Lyotard, 1982, p. 33]. The self, this obstinate impression of a moment of "being-for-itself", of personal identity, each of us claims with self-consciousness as self-consciousness or as competence of self-determination either is seen as an effect of forceful discourse itself or (as in Althusser) as a mere appearance the system necessitates for its functioning. In loving detail, we could draw a picture of this banishment of the self from statements with the same tenor. They show a highly sensible reaction to the modern tendency to over-determination of the self. But at the same time they contort this experience into a founding principle of the discourse they are attached to.

Thesis 4 is to be taken as being in sharpest opposition to this picture of the "self", "I", "Individuality", "Subjectivity"! It stands in permanent opposition in case a linguistically or discursively turned sociology should begin to consecrate "the threatening of subjectivity through the totalitarism of systems of rules and codes" [Frank, 1983, p. 12]. Thesis 4 looks therefore for other principles to organize the relation of *societal reproduction, interaction and individuation* than those of some followers of structuralism, for example.

It puts these three elements into a relation which may be briefly circumscribed by looking at Simmel's second *a priori* [Simmel, 1971, pp. 6-22]. I will call this *apriori* "Simmel's Constellation" as it is a way of defining relations between different components. In my opinion it also marks the place where an explication of the complicated concept of "ground" could start. For I am convinced this constellation points back to that theory of "ground" which Hegel develops in his *Logik der Reflexionsbestimmungen* (book II of *Science of logic*: *Logic of essence and appearance*). It is embedded into his theory of *"Widerspruch"* ("contradiction" is not a good translation; for Hegel is not the one who held combinations of sentences such as "The rose is red *and* the rose is not red" to be reasonable). Again, only a short note on this point is possible: In his famous "excurse" Simmel posed the question: "How is society possible?" The final point of reference for his answer is the "fact of socialization" (*Tatsache der Vergesellschaftung*) (*L*). This fact is said to be the "condition of possibility" of any process of individuation. It is the "Grund" (ground) of any development of a self in persons. But, Simmel says, in the course of this individuation any singular person step by step enters into a double-sided postition. It consists in the fact "the individual is contained in sociation and, at the same time, finds himself confronted by it, ... he exists both for society and for himself" (p. 17). "Simmel's Constellation" concerns a special way of relating the "fact of socialization" with "individuality". The society (*L*) stays the overarching (*"übergreifende"*, Hegel would say) and ultimate condition of the possibility of developing a self. But under its presupposition (together with certain boundary cricumstances, especially in the form of processes of symbolically mediated interaction with significant others) a self is supported (or denied!) *in* or through society — a self which at the same time is exterior to society! This notion of "exterior" indicates possible competences of persons to oppose the conditions of socialization. In other words: Individualization means the development of an autonomous self. But this autonomization — form one point of view — is only possible under societal presuppositions (Mead). On the other hand the development of an *autonomous self* is a decisive condition of the reproduction of society as a whole itself. Insofar persons definitely are *not* seen by Simmel as a mere cross-road for societal or discursive infuences and powers (which they are as well), but as being capable of "beingfor-themselves", if they eventually are able to oppose societal conditions of their own personal existence.

Relating society and individual in a way circumscribed by an decisively anti-hegelian French Haute Couture is, for the moment, fashionable in many places. Simmel's recommendation, however, leads back, in a way, to a nowadays inopportune Hegel. For logically, he says, his constellation emphasizes "the simultaneity of two logically contradictory characterizations of man the characterization which is based on his function as a member, as a product and content of society; and the opposing characterization which is based on his functions as an autonomous being, and which views his life from its own centre and for its own sake" [Simmel, 1971, p. 18].

The logical analysis of this way of relating different components is to be found in Hegel's theory of *Widerspruch*. The paralels of Simmel's formulations to Hegelian statements are evident. To take up only one of the latter: "Als dieses Ganze ist jedes vermittelt *durch sein Anderes* mit sich und *enthält* dasselbe. Aber es ist ferner durch das *Nichtsein seines Andern* mit sich vermittelt; so ist es für sich seiende Einheit und *schließt das Andere aus sich aus*" [Hegel, 1969, p. 49]. Sentences of this type also lead, by complicated paths to the triple meaning of Hegel's also lead, by complicated paths to the triple meaning of Hegel's concept of "ground". 1) "Ground" is seen as an overarching (*übergreifendes*) principle represented in Hegel's Logic by "essence", in Simmel's excurse by the "fact of socialization" (*L*). 2) "Ground" is seen as a movement or process which — on account of its interior contradictions as antagonismus leads to a wreck (Hegel: *Zugrundegehen*) or at least to *Afhebung* as a new constellation on a higher level. 3) The decisive point, however, is that the whole argumentative figure Hegel calls *Widerspruch*[10] is seen as a constellation not to be superseded ("Ground" in the most expanded sense!) he thinks the type of relational structure given therewith a material as well as a fomal arrangement of "moments". His most scandalous statement in this connection reads: "All things are contradictory in themselves" [Hegel, 1979, vol. II, p. 58]. But what he wants to say is that »life« (p. 59) or »Living« is a *processual* arrangement of different (real) components, arranged in a manner which is circumsribed by that way of relating logical elements he calls *Der Widerspruch*: For Hegel something is living (*lebending*) only in so far as it contains the *Widerspruch* and is the power to contain contradiction and to tolerate contradiction in itself. For Hegel, you can't see "behind" this structure, in so far it is *unhintergehbar* as Wittgenstein would say (Ground 3). "Wenn aber ein existierendes nicht in seiner positiven Bestimmung zugleich über seine negative überzugreifen und eine in der anderen festzuhalten, den *Widerspruch* nicht in ihm selbst zu haben vermag, so ist es nicht die lebendige Einheit selbst, nicht Grund (=Grund 3), sondern geht in den *Widerspruch* zugrunde (=Grund 2!)" [Hegel, 1972, vol. II, p. 59]. Simmel's Constellation on the other hand is a way of claiming a material field ordered according to Hegel's principles.

Thesis 4 therefore runs: Thoughts like those in Simmel's second *a priori* give a sketch of the theory of society, person and self the third movement of the pirouette aims at. This type of ordering corresponds to Hegel's figure of *Widerspruch* and means a dialectical configuration of reproduction, interaction and self.

CONCLUDING REMARKS. Building bridges to Hegel's philosophy is treated with the utmost scepticism just now. But I wanted only to provide a mere sketch of the direction I would go in dealing with the Wittgenstein-Problem. The normal philosophical way is to take a linguistic turn and twist individual sentences the analytical style of philosophy. I do not stand in firm opposition to this procedure. There are a lot of problems stemming from everyday-language and specialized terminology. Certainly in my text as well. But I firmly assert that these linguistic turns are not without sociological danger. Where they are imported into sociology through any of the versions of the relation between "language-game" and "life-form" the pirouette to be found in Wittgenstein himself is stimulated again. The special twist Wittgenstein seemingly (I'm sure someone will find the precise quotations negating the problem again) gives to the problem is reflected in controversies in sociology which can be dubbed "social-ontological". What I wanted to show was: In view of the sometimes very sharp contradictions between the three (main?) positions of the pirouette not even the most artful choreography is able to deep all options open. It follows that even a linguistic turn, in any direction, does not spare a sociologist the reflection on the ontology of society he normally takes for granted. Besides, I do not think it is enough to sprinkle "discourse" (in Foucault's coining) or "language-game" (in Wittgenstein's) fashionably through one's won discourse. It happens too often. Contrary to such easy habits, the importation of rather controversial "social ontologies" from philosophy into sociology should quite properly incite some reflections on social ontology in sociology. For the Wittgenstein-Problem is not solely a linguistic problem. Maybe the word "ontology" does not fit. In this case another language-game might be used to pose the problem and deal with it.

Translated by B. Young

Goethe-Universität
Fb 3, Methodologie
Senckenberganlage 15
Frankfurt/M., Germany

NOTES

[1] Is what we call "obeying the rule" something that it would be possible for *one* man to do ...

[2] In Rorty's [1979] opinion this scheme has "considerable plausibility": "The picture of ancient and medieval philosophy as concerned with things, the philosophy of the seventeenth through the nineteenth centuries with ideas, and the enlightened contemporary philosophical science with words has considerable plausibility" (p. 263).

[3] For example "It is my contention that every reality is equally real" [Mehan & Wood, 1975, p. 31].

[4] "The notion of being committed by what I do now to doing something else in the future is identical in form with the connection between a definition and the subsequent use of the world defined which I discussed in the last chapter" [Winch, 1960, p. 50].

[5] Especially in passages where he deals with war; e.g., pp. 130-131.

[6] I call the ethnomethodologists "Serapion's Brethren" on account of the novel of E. T. A. Hoffman *Der Einsiedler Serapion* (Serapion the Anchorite). In this novel all the official principles of ethnomethodology are practised by an anchorite who thinks itself to be identical with one of the ancient Serapions.

[7] The first quotation is from K. Leitner [1980, p. 70], the second from [Mehan & Wood, 1975, p. 190].

[8] Cf. e.g.: "... objective theory in its logical sense (taken universally: science as the totality of predicative theory, of the system of statements meant 'logically' as 'propositions in themselves', 'truth in themselves' and in this sense logically joined) is rooted, grounded in the life world, in the original self-evidence belonging to it" [Husserl, 1970, p. 129].

[9] See for example the respective statements of the scientist E. Jantsch [1982]. He is no Hegelian but defines "spirit" as follows: "In this perspective spirit, appears as dynamic of downright self-organisation" (p. 227).

[10] See his definition of *Widerspruch* [1969, p. 49].

38

REFERENCES

Foucault, M. [1972]. The discourse on language. In M. Foucault. *The Archeology of Knowledge.* New York: Pantheon Books.

Foucault, M. [1978]. *Von der Subversion des Wissens.* Frankfurt/Berlin/Wien.

Foucault, M. [1981]. Das Verschwinden des universellen Intellektuellen. *Frankfurter Rundschau,* v. 27.6.

Frank, M. [1983]. *Was is Neostrukturalismus.* Frankfurt/M.

Gadamer, H. G. [1986]. *Wahrheit und Methode,* Tübingen.

Garfinkel, H. [1967]. *Studies in Ethnomethodology.* Englewood Cliffs: Prentice-Hall.

Gier, N., & Gier, F. [1980]. Wittgenstein and forms of life. *Philosophy of the Social Sciences,* 10, 3.

Hegel, G. W. F. [1969]. *Wissenschaft der Logik,* 2 Bände. Hamburg.

Hegel, G. W. F. [1979]. *Logische Propädeutik, Werke in 20 Bänden.* Frankfurt/M.

Horkheimer, M., & Adorno, T. W. [1979]. *Dialectic of Enlightenment.* London: Verso.

Husserl, E. [1970]. *The Crisis of European Sciences and Transcendental Phenomenology.* Chicago: Northwestern University Press.

Jantsch, E. [1982]. *Die Selbstorganisation des Universums. Vom Urknall zum menschlichen Individuum.* München.

Leiter, K. [1980]. *A Primer of Ethnomethodology.* Oxford.

Lyotard, J. F. [1982]. *Das postmoderne wissen. Ein Bericht.* Bremen.

Mead, G. H. [1934]. *Mind, Self and Society – from the Standpoint of a Social Behaviourist.* Chicago.

Mehan, H., & Wood, H. [1975]. *The Reality of Ethnomethodology.* New York Sydney/Toronto.

Nietzsche, F. [1974]. On truth and falsity in their ultra moral sense. In F. Nietzsche. *The Complete Works* (ed. by O. Levy), vol. 2. New York: The Gordon Press.

Nietzsche, F. [1910]. *The Joyful Wisdom.* London: T. N. Foulis.

Pollner, M. [1970]. In Weingarten et al. (Eds.). *Ethnomethodologie. Beiträge zu einer Soziologie des Alltagshandelns.* Frankfurt/M.

Rorty, S. R. [1979]. *Philosophy and the Mirror of Nature.* Princeton: Princeton University Press.

Searle, J. [1969]. *Speech Acts. An Essay in the Philosophy of Language.* Cambridge: Cambridge University Press.

Simmel, G. [1958]. *Untersuchungen über die Formen der Vergesellschaftung.* Berlin: Dunker und Humblot.

Simmel, G. [1971]. How is society possible. In D. L. Levine. *Georg Simmel: On Individualism and Social Forms.* Chicago: Chicago University Press.

Wieder & Zimmermann [1970]. In Weingarten et al. (Eds.) *Ethnomethodologie. Beiträge zu einer Soziologie des Alltagshandelns.* Frankfurt/M.

Winch, P. [1960]. *The Idea of Social Science.* London: Routledge and Kegan Paul.

Wittgenstein, L. [1976]. *Philosophical Investigations.* London: Oxford University Press.

Wittgenstein, L. [1969]. *On Certainty.* Oxford: Blackwell.

Poznań Studies in the Philosophy
of the Sciences and the Humanities
1991, Vol. 22, pp. 39–57

David Jary

BEYOND OBJECTIVITY AND RELATIVISM: FEYERABEND'S "TWO ARGUMENTATIVE CHAINS" AND SOCIOLOGY

> There is no scientific method ... no single procedure or set of rules ... Every project, every theory, every procedure has to be judged on its own merits and by standards adapted to the procedures with which it deals [Feyerabend, 1975].

> The scientist must appear to the systematic epistemologist as a type of unscrupulous opportunist [Einstein – quoted in Feyerabend, 1975].

1. Introduction

Paul Feyerabend's iconoclastic philosophy with its many racy indictments of "scientific rationalism", though sometimes admired for its satirical bite, is more usually portrayed as promoting a dangerous relativism. Feyerabend protests in vain that it is not reason or "reasonableness" that he wishes to oppose, only abstract versions of Reason; that it is in the present intellectual climate, tipped too far in favour of scientific objectivism, that he stresses "relativistic" elements in science and knowledge.

Feyerabend's accounts of science and his oft-repeated descriptions of his own philosophy as "Millian" (the Mill of *On liberty*) ought to have made clear that his intentions have not been to promote "irrationalism", but rather to advance a proper awareness of the real world complexity of science and its strengths and limitations as a social activity. Along with this would also go an awareness of the nature and needs of human beings. Undoubtedly Feyerabend's self-declared

"Dadaism" and his many polemical fights have mislead many critics into thinking otherwise about the intentions and implications of his thinking (as indeed they were often intended to, by showing how easy it is for the supporters of methodical forms of rationalism to be mislead by their own "rationalism"). Beneath such appearances and such traps, however, Feyerabend's ideas always involve a "serious playfulness", and not at all a onesided celebration of relativism and "irrationalism".

In this paper I want to side with Feyerabend and support his view that a rejection of the stricter forms of scientific rationalism is not only "humane and reasonable" but also good for science. The stimulus to the production of the paper has been provided by Feyerabend's relatively recent provision of a convenient summary formulation of central aspects of his overall position in terms of "two argumentative chains" (one of which is a relativist chain, the other an open, realist chain). These Feyerabend presents as stating ever-present alternatives in social and scientific discourse. I have seized upon this fresh formulation as providing the opportunity to state unequivocally what a careful reading of Feyerabend's work always revealed. This is that, like the work of Richard Bernstein and others (but ultimately in ways which are less loaded and thus more effective in making the general point) the outcome of Feyerabend's philosophy should be seen as carrying us entirely "beyond objectivism or relativism", beyond philosophical dogma. Indeed, the argument I will want to advance is that he might be seen as doing this more effectively than any other philosopher, partly because of his polemic, rhetoric, playfulness, etc., but especially because of his outright refusal to reerect any new kind of permanent system[1]. There are reservations to express about particular aspects of Feyerabend's philosophy — for example, he is on occasions over-dismissive of systematic theory, seeming to reject this even when it does not seek to impose rationalism. The biggest mistake, however, in interpreting Feyerabend's philosophy is to confuse his overall argument "against method", against closure, with his actual (or apparent) advocacy of many avowedly partisan positions, the expression of which his "open" approach allows but does not formally entail.

2. Feyerabend on Popper's falsificationism and "critical rationalism"

Feyerabend's philosophy is post-empiricist, post-positivist, as well as post-falsificationist. But it is, of course, Popperian falsificationism and "critical rationalism" against which Feyerabend has been most outspoken. His best known critiques of Popper appear in two books [1975, 1978]. We will also make reference to volume 1 of his recently published collected essays [1981] and the further collection of papers and essays [1987].

At its simplest Feyerabend rejects Popper's methodological proposals as

> not in agreement with scientific practice, and would destroy science as we know it.

> Even the ingenious attempt of Lakatos to construct a methodology that (a) does not issue orders yet (b) puts restrictions upon our knowledge increasing activities, does not escape this conclusion [Feyerabend, 1975, p. 14].

It is important to see that these conclusions are reached by Feyerabend from an historical, psychological, and (to an extent) a sociological analysis of science. Equally important, they also arise from a careful consideration of the examples of successful science that are actually advanced by Popper and Lakatos.

What soon becomes apparent is that science is successful in ways which are not in accord with Popper's methodology. Falsificationism (whether in its "naive" or its "sophisticated" versions) is *not* the "solution to the problem of empiricism" which Popper suggested it to be. The relative autonomy of the "facts" as required by falsificationism proves unjustified. As Feyerabend puts it:

> experimental results, 'factual' statements, either contain theoretical assumptions or assert them by the manner in which they are used [Feyerabend, 1975, p. 31].

Thus there can be no strictly theory-independent test of theories of the kind that falsificationism clearly requires[2].

That a circularity exists in the formulation and testing of theories in this way also means that the innovative (in Kuhn's terms "revolutionary") science which Popper himself so much admires, cannot be falsificationist. Instead,

> The first step in our criticism of familiar concepts and procedures, the first step in our criticism of "facts", must be an attempt to break the circle. We must invent a new conceptual system that suspends, or clashes with the most carefully established observational results, confounds the most plausible theoretical principles, and introduces perceptions that cannot form part of the existing perceptual world [Feyerabend, 1975, p. 32].

A famous example given by Feyerabend is the way in which Galileo set about establishing his own heliocentric view. This was *not* falsificationist. Galileo's theories flew in the face of "commonsense". Furthermore, his telescopes were poor. What was "visible" through these depended on which theories one already accepted and on what one expected to see. The predictions that Galileo was at first able to make were more limited, and of far less practical use (in navigation, etc.), than those possible using the earlier epicyclical theories. The decisive point against falsificationism is that anyone wanting to refute Galileo's theory using the "facts" could readily have claimed to have done so. It is in such circumstances that an innovative scientist like Galileo will sometimes need to be "opportunistic", use persuasive tricks, and so on, in establishing any new view.

Feyerabend is convinced that "progress" in science would have been blocked if Galileo in particular, or science in general, were actually to have adopted falsificationism as its only method. Feyerabend declares:

> To ask how one will judge and choose in as yet unknown surroundings makes as much sense as to ask what measuring instruments one will use on an as yet unknown planet [Feyerabend, 1978, p. 29].

There is, he insists:

> no (falsificationist) decision but a natural development leading to traditions which in retrospect give reasons for the action had it been a decision in accordance with standards [Feyerabend, 1978, p. 29].

Even where agreement exists on a potentially refuting instance, a single "refuting instance" is rarely decisive. There are almost always unexplained "anomalies" which might refute a theory. For example, Mercury's perihelion was long known to be other than what was predicted by Newton's mechanics — potentially a refuting instance. But his was not taken as refuting Newton's theory. Instead it remained simply an unexplained anomaly until a more acceptable overal theory-and-facts was provided by Einstein.

The awkward question that obviously arises about falsificationism is "Would anything falsify falsificationism?" One commentator [Williams, 1975] has even described Popper as the "mad mechanic" of the philosophy of science, struggling by ad hocery to keep "his old banger" on the road. The effects of Lakatos to save falsificationism occur in this context, but shifting the ground to talk of "sophisticated falsificationism" and "progressive and degenerating scientific research programmes" (in which a single refuting instance is no longer decisive) only compounds the difficulties for falsificationism. It succeeds only in emphasising that falsificationism lacks a decisive cutting-edge; and it can do little either to save the idea of a methodologically principled critical rationalism, whether on Popper's or on Lakatos' terms. Feyerabend's typically "playful" summing up of the situation is instructive both in relation to Lakatos' and his own methodological position. Feyerabend insists that he never used the catchphrase "anything goes" to summarise his own view. Rather he introduced this to sum up what he saw as the only outcome of Lakatos' attempt to save a single method. If Feyerabend does sometimes describe his own attitude to method, in contradistinction to Popper's, as "anarchistic", what Feyerabend "wickedly" suggests about Lakatos is that he is "an anarchist in disguise"[3].

3. The importance of the "paradigm" concept in Feyerabend's philosophy

What Lakatos' concept of competing "scientific research programmes" recognises, but in the end fails to follow through to its fullest implications, is the utter centrality of particular traditions or "paradigms" in science, each with their own *particular methods*. Many philosophers, historians and sociologists of science apart from Feyerabend have nowadays come to acknowledge the importance of "scientific paradigms" in this way, notably, of course, Thomas Kuhn. However, it is the particular way in which Feyerabend deploys the paradigm concept which I will want to show as at the heart of the strengths of his overall viewpoint, but which also leads to many misinterpretations of its implications.

In view of all the special claims made for the "scientific method", it is somewhat ironic that it is "hermeneutics" and the "interpretive understanding" of particular scientific paradigms and traditions which Feyerabend regards as basic in providing an understanding of any scientific knowledge. The concept of the scientific paradigm, of course, was originally Thomas Kuhn's, for whom a scientific paradigm is:

> a universally recognised scientific achievement that *for a time* provides model problems and solutions for a *community* of practitioners [Kuhn, 1970, viii, ital. added].

For Feyerabend, as for Kuhn, the great advantage of the concept of paradigm, compared with Popper's or any similar philosophical formalism, is its reference to a "flesh and blood" and a psychological and sociological reality, to scientific groupings, to "traditions", to social activities, and the like. This being so, the ambiguities and open-endedness of the paradigm concept, which some commentators suggest undermine its usefulness, must be seen as a virtue. Within Feyerabend's work the concept of "form of life", taken from Wittgenstein's more sociological "second philosophy", is additionally significant. Again, it is the diffuseness and suggestiveness of this concept that is of value.

A number of well-established features of both Kuhn's and Feyerabend's account of science follow from the centrality they accord paradigms within science. Among these are the fact that a scientific paradigm is not itself so much judged or tested but is the basis of such judgements; that in the absence of any theory-neutral data language, the conceptual and logical relations between paradigms will often be "disjoint", preventing any simple comparisons of paradigms; that scientific beliefs and changes in these involve psychological attachment to groups (including "gestalt" switches), which are not dissimilar from our other beliefs, for example, about politics and about ways of living in general; that "progress" in science is thus "discontinuist", involving "revolutionary breaks"; that new paradigms do not usually fully incorporate old knowledge; and that something ontologically or empirically useful in earlier world views often gets lost, much as, say, earlier craft knowledge gets lost.

The polemical way of summarising all of this is, of course, that the relation between alternative theories or competing paradigms in science will often involve elements of incommensurability[4]. Since it is here that the critical attack and problems with the interpretation of Feyerabend (as well as Kuhn) particularly arise, it is also vital to stress what the two theorists *do not* say. Above all, they do not present the presence of paradigms and incommensurability in science as undermining science. On the contrary, the effectiveness of paradigms in creating new knowledge and making visible "entirely new worlds" is strongly stated. But what Feyerabend in particular does do (and here is one important difference from Kuhn) is to point out the limits of science, especially in its institutionalised forms. Is is equally important to recognise, however, that this is a different matter from "relativism" or "irrationalism". Yet it is plain from the reception of his work that Feyerabend does not convince his critics of this; that his ideas frighten some and anger others.

4. Feyerabend's "relativism": the reception and misrepresentation of Feyerabend's work

The charge that Feyerabend's work involves a dangerous relativism and that it sanctions "irrationalism" is widespread. It is heard from commentators as diverse as the sociologist Anthony Giddens [1980], the realist philosopher Roy Bhaskar [1979] and many more[5]. Noting the "playfulness" of much of Feyerabend's writing, some commentators even suggest that he is not a serious philosopher. Steven Rose [1979], for instance, notes contemptuously that on the dust cover of [1978] Feyerabend would appear to be wearing a "party hat"! Undoubtedly Feyerabend's characterisation of his approach to philosophy as "anarchistic" or Dadaistic has encouraged views of this sort. However, he is insistent that such phrases do not represent a well-rounded view of his philosophy.

Feyerabend's response to his critics is scathing. The titles he gives to his replies to his Popperian critics — for example, "Conversations with illiterates", "Marxist fairy tales from Australia" [Feyerabend, 1978] — indicate his view of his self-styled rationalist opponents, critics who do not read carefully and who often rush to rationalist recipe judgements faced by the deliberate traps that he sets for them[6]. As he sees it, the charge of "relativism" is little more than a "craving for intellectual security", the arrogant defensive reflex of those who assume that the *only* choice is between irrationalism and their own philosophy![7]

Feyerabend is adamant that he is not a relativist in any ultimate sense of the term:

> Philosophical relativism is the doctrine that all traditions, theories, ideas, are equally true or equally false, in even more radical formulation, that any distribution of truth values over traditions is acceptable [Feyerabend, 1978, p. 83].

This form of relativism, Feyerabend insists, is nowhere defended in his work. Rather his relativism is:

> of precisely the kind that seems to have been defended by Protagoras ... [Feyerabend, 1978, p. 82].

It is civilised and reasonable:

> because it pays attention to the pluralism of traditions and values
> [Feyerabend, 1978, p. 28].

In science it allows for the incorporation of elements of earlier "rejected" theories. In social life it allows profit from the study of alien cultures. And it allows members of minority traditions of all kinds to either protect their own world views or to move beyond them. It is in the spirit of doing justice to complexity that Feyerabend acknowledges the possible merit of traditions rejected by western science, such as "folk" medicine, for example. Thus it is that Feyerabend's opposition to orthodox methodology, as well as giving proper respect to particular paradigms, particular forms of life, also stands for "learning from the comparison of practices", and for "dialectical growth of knowledge", for "reason" as well as "experience".

Though he draws on Wittgenstein (and also Whorf), incommensurability is for Feyerabend only one possible relation between paradigms, not a necessary one. Incommensurability between paradigms (in Wittgensteinian terms, incommensurable "forms of life"), and also potential commensurability and the possibility of scientific "progress" are equally acknowledged, though not, of course, in a falsificationist way. It is clear that Feyerabend's use of Wittgenstein involves a markedly different reading of this philosophy than some others. Illustrative here is Gellner's [1974] sweeping dismissal of Wittgenstein and Winch (a dismissal he also extends to Feyerabend — [Gellner, 1975]), as follows:

WM → ISS

But ISS is absurd

.:WM is absurd

(where WM = Wittgenstein's mature philosophy and ISS = Winch's *The idea of a social science*)!

5. Feyerabend's Millian "method"

Feyerabend's basic point showing through all the rhetoric, and whatever the appearance of irrationalism, is that ultimately if we want progress in science we must keep our options open:

the world we want to explore is a largely unknown entity. We must not restrict ourselves in advance [Feyerabend, 1975, p. 20].

Dogmatic epistemology, Feyerabend suggests, "maims by compression, like a Chinese lady's foot". It is clear, Feyerabend continues, that, as well as resting on a "craving for intellectual security":

> the idea of a fixed method ... rests on too naive a view of man and his social surroundings [Feyerabend, 1975, p. 27].

A Popperian debate is seen by Feyerabend as a "guided exchange". In:

> an open exchange, on the other hand, ... (the) tradition (or philosophy) adopted by the parties is unspecified in the beginning and develops as the exchange goes along. The participants get immersed into each others' ways of thinking, feeling, perceiving, ... world views may be entirely changed ... [Feyerabend, 1978, p. 29].

Rather than empiricism, positivism, falsificationism or any dogmatic view, including outright relativism, what Feyerabend's philosophy ultimately involves is an open discourse theory, or, as he himself expresses it, a Millian approach:

> The attempt to increase liberty, to lead a full and a rewarding life, and the corresponding attempt to discover the secrets of nature and of man entails ... the rejection of all universal standards and all rigid traditions [Feyerabend, 1975, p. 20].

And for Feyerabend:

> An account and a truely humanitarian defence of this position can be found in J. S. Mill's On liberty. Popper's philosophy, which some people would like to lay on us as the one and only humanitarian rationalism in existence today, is but a pale reflection of Mill [Feyerabend, 1975, p. 98].

Compared with Mill's approach, Feyerabend suggests:

> Popper's philosophy ... is much more formalistic and elitist [Feyerabend, 1975, p. 48].

It is "liberty" that allows questions about the "interests" served by particular scientific theories to be raised as part of the relevant desiderata in the appraisal of science. It is "liberty" which allows the church-like status of science — what Illich calls the "radical monopoly" of science — to be continuously challenged (though, if the rule of experts is thus rejected, the advice of the experts is a different matter). "What's so great about science?", asks Feyerabend at one point. Its perspectives are often partial, its particular applications have not always advanced humanity. This is the reason why Feyerabend also suggests that science, if defined in particular terms, should be seen as "one tradition among many", and judgements about knowledge, which concern us all, should be open to all, unhindered by any dogmatic method. It is Millian proposals which provide protection against what Feyerabend calls the "way of power".

In his most recent writings Feyerabend has elaborated some of the many reasons for supporting a Millian view. Among these are: that no-one is infallible; that any position is always likely to "contain a portion of the truth", that any position held without challenge will be held as a "prejudice"; that contrasting views will help clarify a particular viewpoint; that human ingenuity is not limited [Feyerabend, 1987, p. 34]. However, I would want to argue that it does not matter that the formal conditions for "liberty" are *not* settled, or even if these are "inherently irresolvable", "essentially contested", etc. For this can be seen as reinforcing the argument for "open method". As Feyerabend himself says, none of his detailed arguments for a Millian view are essential, any of them might be changed without undermining the argument "open method". This is also among the reasons why Feyerabend's Millian approach should not be seen as the erection of a "new general method".

It is because they are more elaborated than Feyerabend's that I would suggest that comparable models, such as Habermas' model of "open discourse", prove less satisfactory. For Habermas the "will to truth" is implicit in the very structure of the "speech act":

> experience supports the truth claim(s) of assertion ... but a claim can be redeemed only through argumentation [Habermas, 1976, xvi].

> Truth is not the fact that a consensus is realised but rather that at all times at any place, if we enter into a discussion a consensus can be realised under conditions that recognise it as a justified consensus. "Truth" means warranted assertability. "Truth" cannot be analysed independently of "freedom" and "justice" [Habermas, 1976, xvi].

For Feyerabend:

> The debates settling the structures of a free society are open debates ... a free society will not be imposed but will emerge only where people solve particu-

lar problems in a spirit of collaboration ... a free society is a society in which all traditions are given equal rights, equal access to education and other positions of power [Feyerabend, 1978, p. 30].

And for Habermas:

all participants must have the same chance to initiate and perpetuate discourse, ... to give reasons for and against statements, to express attitudes, feelings, intentions and the like [Habermas, 1976, XVII].

Compared with Feyerabend, however, Habermas' conception of truth rests on a general theory of "human interests" and "communicative competence" the detail of which, though much admired, is highly contentious. And Habermas has found it necessary to continuously revise his intricate and ingenious "system", despite once claiming its "quasi-transcendental" status. None of this is to deny that Habermas' sociology is not of great value, not only in its theoretical analysis of "human communication", but also for its systematic analysis of the power structures of modern society. These are areas where Feyerabend's analysis is only polemical and perfunctory. The point remains that it is Feyerabend's "open-ended" justification of "liberty", almost banal in comparison, that in the end serves best as a general framework.

6. Feyerabend's "two argumentative chains"

It is in his collected essays that I want to show Feyerabend finally removing any excuse for continued misunderstanding of his view. It is here that, in the space of a few pages, he provides a succinct summary of his overall position in terms of "two argumentative chains". Through these he makes clear that neither rule governed objectivism *nor* relativism can be sustained as a sole general position.

In the first of these chains [Feyerabend, 1981, p. xiii], "accepting a form of life L we reject a universal criticism and the realistic interpretation of theories not in agreement with L":

$$L \text{ (FORM OF LIFE)} \rightarrow \overline{\text{CRITICISM}} \rightarrow \overline{\text{REALISM}}_L$$

This chain represents the closed world of the paradigm and the "virtual incommensurability" which may exist between paradigms. It is also the position of any other "closed" form of life, for example, any closed society or subculture, which, as Feyerabend suggests, may not regard the position as a "philosophy of defeat" but as a life enhancing ethical and political stance. As Alisdair MacIntyre [1981] asserts "what matters at this stage is the construction of local forms of community within which civility and moral life can be sustained" (p. 244). An isomorphism obviously exists between the advantages of retaining a successful paradigm in Kuhnian normal science and what can happen in such a "closed" culture.

In the second chain, however, "criticism" means [Feyerabend, 1981, p. ix] that, "we do not simply accept the phenomenon, processes, institutions that surround us but we examine these and try to change them":

CRITICISM → PROLIFERATION → REALISM

"Proliferation" means that "we use a plurality of theories (systems of thought, institutional frameworks) from the very beginning." Unlike the situation within a particular tradition or in Kuhnian normal science:

> We do not work with a single theory, system of thought, institutional framework until circumstances force us to modify it or to give it up [Feyerabend, 1981, p. ix].

The arrows in this model do not "express a well-defined connection such as logical implication", but rather:

> that starting from the left hand side and adding physical principles, psychological assumptions, plausible cosmological conjectures, absurd guesses and plain common-sense views, a dialectical debate will eventually arrive at the right hand side [Feyerabend, 1981, p. x].

It is in terms of such a chain that growth or "progress" in knowledge trascending a particular paradigm can be stated, and this would include Kuhn's "revolutionary science". Once established, of course, such a "revolutionary science" becomes "normal science" — a return to the first chain. But the possibility of a further transcending of any dominant normal science paradigm or a coalescence of views remains. It is the ever present possibility of these two chains which corresponds to what Kuhn calls the "essential tension between tradition and innovation" in science [Kuhn, 1977]. It is also the basis on which MacIntyre's moral community can expand its horizons from a local to a wider context, however difficult this may sometimes prove. What the existence of two chains also finally means is that "science" can variously be seen as either a distinct form of life or as continuous with other forms of life; that "science" is not only like politics or religion, as Kuhn suggests, but also part and parcel of these.

7. Feyerabend's methodological strictures and sociology's "pluralism" as consonant: an "open framework" for sociology?

As a sociologist, what I want finally to suggest about Feyerabend's treatment of method is that the framework of two argumentative chains and its Millian basis is particularly well-placed to provide a "rationale" for sociology's existing pluralism. Another way of making the same point is to say that sociology's existing practice and Feyerabend's strictures are broadly consonant. Where they are not (and particular methods are still asserted as universal methods) my argument is that it would be better if sociology's pluralism and Feyerabend's strictures were accepted. In making this suggestion I make no assertion that such an application of Feyerabend's ideas would actually be sanctioned by him. The likelihood is that they would not, for two main reasons. First, Feyerabend's *forté* is iconoclasm and critique, not construction. Second, despite the sociological influences on his thinking, Feyerabend has directed no attention to examining sociology in-the-round, and his relatively few comments on the subject have mostly been to express sceptical dismissals of the scientistic pretensions of some of its practitioners. Nonetheless, I do maintain that a rationale for sociology's present condition can be expressed in terms of Feyerabend's framework, even if Feyerabend himself might be expected to have some reservations about so overtly "constructive" a use of his thinking.

A first obvious attraction of Feyerabend's overall position is that it accommodates (and could domesticate) the long but often troublesome tradition in sociology of analysis, which is made on the one hand in terms of actors' own categories, and on the other hand in terms of general sociological concepts and theories which move beyond actor's categories. While disputes about the relative merits and appropriateness of these two approaches have been deeply divisive, a model of two chains fits an increasing *detente* within the subject, in which both modes of analysis are recognised as valid. Feyerabend's rejection of ontological or epistemological dogmas of any kind would also tame the disabling influences of doctrines such as phenomenology, with their suggestion that large areas of sociological research must either await the solution to ontological problems or for ever founder on insoluble difficulties such as "indexicality"[8].

A second aspect of sociology where Feyerabend's thinking and sociology's practice are in accord is that Feyerabend's framework is well placed to accommodate the great variety of schools of thought and particular approaches within the subject. It can accomplish this while avoiding any assumption that such a pluralism must involve warring factions. But it also accomplishes this without assuming that proliferation is only justified if it is *en route* to an ultimate unitary paradigm. Kuhn's conception of a "scientific" discipline is that it must always have as its objective an agreed paradigm — as Bailey [1980] puts it, for Kuhn a proliferation of perspectives is always a "strategic" matter. For Feyerabend, in contrast, the implications of proliferation (although they may sometimes be "strategic") are much more open-ended. Accepting Feyerabend's framework and expressing sociology's practice in terms of it would at once be more optimistic about the possibility of "progress" within the discipline than a "warring camps" view, and less "optimistic" than any view which expects ultimate paradigm integration. Since those proposing a unitary goal for the discipline often acknowledge that the present state of sociology is at best "pre-paradigmatic" or even in its "infancy", a Feyerabendesque account of sociology might also be regarded as more satisfactory in registering the existing strengths and achievements of the subject in capturing a complex social world that is best viewed pluralistically[9].

A third way in which Feyerabend's framework is potentially useful is in its capacity to provide a general vindication of sociology's continuing constitution as a "broad" academic discipline, a discipline relatively unconstrained by conventional disciplinary boundaries, possessing multiple goals, and with interests in what Bailey [1980] refers to as "human survival" and "sensibility" as well as in narrower forms of scientific truth. This is obviously a different matter from providing a justification for those more "utopian" expressions of the same general orientation (e. g., some versions of Marxism or critical theory) where the goal is to collapse such analytical and ethical general goals into one true "theoretical practice". In a Feyerabendesque account of sociology, such "totalising" approaches can only receive justification as particular approaches; they must take their place within a broader "critical cultural discourse" [Gouldner, 1977a], of which even sociology is only a part[10].

There is one further way in which I see Feyerabend's framework as potentially of valuable. This is the flexibility it can allow in dealing with issues of "facts and value". Obviously invaluable in making clear how social standards and social interests enter the process of research and discovery, Feyerabend's formulation is also helpful in two further ways. The first is in making plain the terms in which actors' and analysts' meanings might be merged in reaching accounts of the "facts" or of "values" (though Habermas does this more fully). The second is in doing so without imposing any requirement to operate with the kind of dogmatic ontological distinction between "facts" and "values" that is sometimes found in sociology, in which values are left entirely within the realm of the "non-rational". Ironically, in view of all the accusations directed at Feyerabend, a dogmatic assertion that values are non-rational is often a feature of the thinking of many who would otherwise pride themselves on their rationalism, notably Max Weber, and more recently Steven Lukes [1974][11]. I do not deny that Feyerabend is sometimes to be found supporting positions that emphasise the difficulties of any coalescence of "values", as well as the advantages in some situations of not seeking this. (Indeed, we have seen that it is an advantage of Feyerabend's position that it does do this). But his general position is always to provide for a possible coalescence.

In suggesting that Feyerabend's arguments and sociology's best practice are consonant, I should make clear that it has not been my intention to suggest that Feyerabend provides an entirely *new* method. That suggestion would be completely foreign to Feyerabend's own orientation, which is to leave practices "free" to develop in their own way so long as they do not do this by crushing dogmatically other methodological practices. (This also means that none of the particular ways in which I have suggested that a consonance exists are to be seen as *essential* arguments about how sociology must be.). Nor do I make any suggestion that Feyerabend's works are the place to turn to for systematic development of the implications of his general position. For such elaborations there are others, like Habermas, who must be consulted. For reasons of temperament, as well as his views on the existing balance between the forces of restriction and the forces of openness, Feyerabend has not chosen to develop his ideas in a sociological direction. Doubtless his own personal preference, like that of the mature Wittgenstein, is to celebrate diversity. Thus while some of his more "outrageous" statements in favour of minority positions are

54

traps for rationalists, others are from the heart. But his overall position is to not preclude general construction, in social science or elsewhere. A sociology that seeks generalities while also preserving an awareness of diversity and complexity is what I would see as the fruitful outcome of any critical absorbtion of such an amalgam of ideas.

Conclusions

In concluding this paper, there is one further complaint about Feyerabend's thinking which must be dealt with. The complaint is from Richard Bernstein whose important book on these issues has as its title *Beyond objectivism and relativism*, a slogan I have also used in the title of this paper. Bernstein's criticism of Feyerabend is that he fails to offer any convincing account of "good reasons" for beliefs — the only way Bernstein sees of saving rationality. His verdict on Feyerabend is that he is a satirist and the goal of satire is "to ridicule, root out, exorcise" (p. 63). Exactly so. But my argument has been that Feyerabend is much more than this. Much as Bernstein speaks of Kuhn:

> his intention has not been to claim that scientific inquiry is irrational but ... to show the way to a more open, flexible, and historically oriented understanding of scientific activity *as rational* [Bernstein, 1983; my ital.].

Except that Feyerabend, wishing to completely exorcise the old ways, would take more care than Bernstein, lest in asserting a particular formula for rationality one also reinstates dogmatic methodology, even a doctrine of rationality as relatively benign and open-ended as the "practical discourse" that Bernstein is at such pains to elaborate. Given this, my own view is that there is no barrier within Feyerabend's position to accommodating accounts of "good reasons", but these must be introduced without any pretence that there exists any special recipe. This then is where Feyerabend scores, where satire and iconoclasm pay-off, but where Bernstein's more anxiously constructive view (however valuable in other ways) risks leading back to dogmatism.

And so Feyerabend's central point remains: that there exists no final rule of method, no single identifiable basis of rationality. The "rationalists" are wrong to suggest otherwise. It is in this context that Feyerabend refers to the philosophy of science as "a subject with a great past"! [Feyerabend, 1956]. On too many occasions — one could cite Comte, Logical Positivism, Popper, and many more — it has had as its goal the breaking free of dogma, only to slip back, betraying its own claims to be a "critical philosophy of science". In this sense, the verdict on Popper's "critical rationalism" would be "not critical enough". What all Feyerabend's polemic and iconoclasm is designed to avoid is any further slipping back on these issues. But I hope that I have shown that it does this without ushering in irrationalism; that his position truly carries us beyond objectivism or relativism.

Department of Sociology
Staffordshire Polytechnic
Leek Road, Stoke—on—Trent, ST 4 2 DF
United Kingdom

NOTES

[1]Another strong statement of a non-dogmatic approach to methological and ontological issues which might do much the same job is Rorty [1983], though no attempt is made to discuss this here. It should be noted that I am not a philosopher. My valuation of Feyerabend's philosophy will be very much from the point of view of a sociologist rather than always in ways appropriate to more technical philosophy. There are many technical achievements of philosophy that, like Feyerabend himself, I would not wish to disparage, while nonetheless insisting on escape from what Rorty calls the "Cartesian anxieties" of professional philosophers.

[2]Popper acknowledges the theory-relativity of "facts" but fails to see the implications this has for falsificationism.

[3]At LSE, where both Lakatos and Feyerabend worked, Lakatos would from time to time burst in on Feyerabend's lectures from an adjacent room to contest a particularly contentious point. It helps in *Against method* in its proper context to know that much of the *raison d'être* of this book was to provoke a response from Lakatos. Unfortunately Lakatos died before this was possible.

[4] "Incommensurability" is sometimes suggested, but wrongly, as a matter of strict logical incompatibility; that would require common concepts. Although on some occasions in his early papers Feyerabend does appear to argue such a logical incompatibility (as he readily concedes), it is clear that any such claim would be directly at odds with the general thrust of his writings, where the claim is that common concepts are lacking.

[5]Among the further long list of critics of Feyerabend's views we can note: Agassi, Macham, Newton-Smith, Stove, and Watkins.

[6]An example of such recipe judgements provided by Feyerabend is of a large number of rationalist thinkers who attached their names to a manifesto condemning astrology, only to decline an invitation to air their viewpoint on radio, on the grounds that they knew little about the subject! Here as elsewhere, Feyerabend's point (misunderstood by most of his critics) is not so much to vindicate astrology or similar forms of life but to illustrate

56

how frequntly "rationalism" involves recipe judgements. The frequent difficulties in distinguishing between "science" and "non-science" are further illustrated by the fact that scientific establishments must often resort to vigilante tactics as in the case of the US Commitee for the Investigation of the Paranormal or, in Britain recently the affair of Benveniste.

[7]I would agree with Feyerabend and other commentators who suggest that the massive public aclaim for Popper's work is a topic in its own right – "the Popper phenomenon", as Mellor [1977] refers to it. The anxious excitement surrounding Feyerabend's work is equally a topic in this right, fully justifying the way in which in *Science in a free society* Feyerabend makes the reception of his own ideas a case study in the limitations of "rationalist" methodology.

[8]For example, Anderson *et al.* [1987], writing from a phenomenological and ethnomethodological perspective voice outright scepticism concerning the ability of any "strong programme" in the social studies of science to achieve its objective of a causal account of science. Although these authors succeed in making many telling points about the difficulties involved this cannot hide that in their end their efforts to exclude an entire approach rest on dogma.

[9]As I see it, the existence of particular perspectives in sociology is not simply a matter of competing "world views", but also arises from a focus on different levels (e.g., micro/macro) and different aspects of reality (e.g., person, group, global structure).

[10]Gouldner's remarks are particularly of interest because his concept of "critical cultural discourse" rests partly on Feyerabend's ideas. The concept also plays a role in his contention that ruling classes are always "flawed" – see Gouldner [1977].

[11]A Feyerabendian position on "inherently contested concepts" in social science would be much closer to Gallie's original, more open-ended and non-dogmatic formulation of this concept – see Gallie [1956].

REFERENCES

Anderson, R., Hughes, J. & Sharraock, W. [1987]. Some initial difficulties with the sociology of knowledge – a preliminary examination of the "strong programme". *Occasional Papers in Social Science*. Manchester: Department of Sociology, Manchester Polytechnic.

Bailey, J. [1980]. *Ideas and Intervention – Social Theory for Practice*. London: Routledge and Kegan Paul.

Bernstein, R. [1983]. *Beyond Objectivism and Relativism*. Oxford: Blackwell.

Bhaskar, R. [1979]. *The Possibility of Naturalism*. Brighton: Harvester.

Feyerabend, P. [1956]. Philosophy of science: a subject with a great past. In R. Stuewer (Ed.). *Minnesota Studies in the Philosophy of Science* (pp. 172-183). Minnesota: Minnesota University Press.

Feyerabend, P. [1975]. *Against Method*. London: New Left Books.

Feyerabend, P. [1978]. *Science in a Free Society*. London: New Left Books.

Feyerabend, P. [1981]. *Realism, Rationalism and Scientific Method*. Cambridge: Cambridge Univ. Press.

Feyerabend, P. [1987]. *Farewell to Reason*. London: Verso.

Gallie, A. [1956]. Essentially contested concepts. In *Proceedings of the Aristotelian Society* (*pp. 167-98*), **56**.

Gellner, E. [1974]. The new idealism – cause and meaning in the social sciences. In A. Giddens (Ed.). *Positivism and Sociology*. London: Heinemann.

Gellner, E. [1975]. Beyond truth and falsity. *British Journal for the Philosophy of Science*, **22**, 331-42.

Giddens, A. [1980]. Positivism and its critics. In T. Bottomore & R. Nisbet (Eds.]. *A History of Sociological Analysis*. London: Heinemann.

Gouldner, A. [1976]. *The Dialectic of Ideology and Technology*. London: Heinemann.

Gouldner, A. [1976]. *The Future of the Intellectuals and the Rise of the New Class*. New York: Seaburt Press.

Habermas, J. [1976]. *Legitimation Crisis*. London: Heinemann.

Kuhn, T. [1962]. *The Structure of Scientific Revolutions*. Chicago: University of Chicago Press.

Kuhn, T. [1977]. *The Essential Tension*. Chicago: University of Chicago Press.

Lakatos, I. [1970]. Falsification and methodology in scientific research programmes. In I. Lakatos & A. Musgrave. *Criticism and the Growth of Knowledge*. Cambridge: Cambridge Univ. Press.

Lukes, S. [1974]. *Power – a radical view*. London: Macmillan.

MacIntyre, A. [1981]. *After Virtue: A Study of Moral Theory*. Notre Dame: University of Notre Dame Press.

Mellor, D. [1977]. The Popper phenomenon. *Philosophy*, **52**, 200.

Popper, K. [1959]. *The logic of scientific discovery*. London: Hutchinson.

Rorty, R. [1983]. *Philosophy and the mirror of nature*. Oxford: Blackwell.

Rose, S. [1979]. If anything goes, so what. *Guardian*, 28 Feb.

Williams, K. [1975]. Facing reality – a critique of Karl Popper's empiricism. *Economy and Society*, 309-358.

Winch, P. [1958]. *The Idea of a Social Science*. London: Routledge and Kegan Paul.

Poznań Studies in the Philosophy
of the Sciences and the Humanities
1991, Vol. 22, pp. 59-85

Leszek Nowak

THE DEFENSE OF A SOCIAL SYSTEM
AGAINST ITS IDEOLOGY
A case study

I.

1. The program of sociology of knowledge differs a great deal from that of the theory of social consciousness. The task of the latter domain is to find an answer to the question of how a given doctrine spreads in a given society. Instead, the program of sociology of knowledge is essentially epistemological — it tries to find dependencies between the social conditions and the truth-values of the appropriate doctrines.

The theory of social consciousness deals with the factual relationships of selection of ideas and their back influence upon the state of social system. The classic writings in sociology of knowledge — those of Marx or Mannheim [1937] — deal with epistemological ones trying to relate the truth-values of a theory to the historical conditions. The differences between different approaches to that domain begin when one asks of the sense of the notion of "truth" or the "historical conditions".

2. As the present paper belongs to the so-conceived sociology of knowledge[1], it presupposes two logically prior theories — a certain epistemological conception of truth [Nowakowa 1976, 1977, my 1977 and 1980] and a certain theory of historical process [my 1979, 1983, 1987]. Its main task is to apply the general claim of sociology of knowledge, the claim of the Marxian origin, to Marxism itself asking a question of how the truth-value of this theory was changing along the historical development of the system to which it has been attached as its official ideology. It is obvious that such a task requires to hypothetically adopt a certain image of that system as the true image of the system itself. Therefore, rather strong assumptions must be accepted to make the setting of Marxism with the system it functions in possible. The epistemological assumptions are shortly presented, together with the interpretation in question itself, in part II of the paper. In part III the assumed theory of the historical process is shortly summarized — in a fragment relevant for the present paper. Based on these assumptions, in part IV truth-content of Marxism is analysed as being dependent on the historical development of socialist historical formation. And in part V some more general conclusions — mainly concerning the structure of ideology and utopia — are drawn from our case study.

II.

1. The first thing that requires to be somehow clarified is the way the notion of Marxism is to be understood. It seems to be useful to distinguish two notions of Marxism. According to the narrow one, it is composed of three parts: (c) Marxian historical materialism, that is, a theory claiming the priority of the economic base to the political superstructure and to the social consciousness and maintaining at the same time that it is the class struggle which plays the leading role in transitions from one mode of production to another; (cc) the doctrine of the liberating mission of the working class, that is, a claim that actually this class is the first class of direct producers which is able to liquidate the foundation of any existing exploitation viz. the private property of means of production and to introduce, with the aid of the worker's state, socialist relations of production; (ccc) the axiology of revolutionary struggle, that is, a doctrine maintaining that for building a new society it is necessary to spread a pattern of revolutionary, i.e. somebody who consciously and of the moral vocation participates in the class struggle.

As is seen, Marxism in the narrow sense contains the social philosophy of the old Marx. Of course, it is a highly controversial decision as a lot of interpretations of Marxism exist that are based on the writing of young Marx, e.g. Kolakowski [1978]. I shall apply further the term "Marxism" only in that narrow meaning.

The so understood Marxism presupposes definite philosophico-methodological background; these elements, taken together, form Marxism in the larger sense. Who accepts Marxist social theory must accept the dialectical philosophy as well. But not the reverse so. It is possible to accept philosophical assumptions of the old Marx's social theory not having accepted the latter at all, that is, claiming that the theory is a faulty realisation of those assumptions. It is what the present author thinks to hold. I accept the dialectical philosophy (taken in some interpretation, called idealisational-categorical [cf. 1977a-b] but it seems to me that Marxist social theory is an erroneous application of it as leading to so-called paradox of historicism [cf. 1979].

2. Of the mentioned dialectical assumptions one is to be shortly outlined here. According to Marx's concept of abstraction (taken in the interpretation that is adopted here) scientific laws are idealizational statements in the form of counterfactual conditionals. They possess in their antecedents idealizing conditions neglecting what is considered to be secondary factors, while in their consequents relationships between investigated magnitudes and factors treated as principal for them are attempted to be discovered. For such an idealizational statement the classic concept of truth is being useless as it is in principle emptily satisfied and the idea of "truth through caricature" seems to be more plausible [cf. 1977c]. According to it, truth is attainable not through representation of a phenomenon but through its deformation decomposing it into its principal and secondary determinants. Roughly, a given idealizational statement is the more (essentially) true, the closer is the image of the essential structure of the phenomenon in question it presupposes to the essential structure of that phenomenon itself. If the image is identical with the structure, that is, if recognizes all the principal and all the secondary factors, then the given statement is absolutely true. If the image recognizes properly only principal factors, and all of them, but not all the secondary ones, then the appropriate statement is termed relatively true. If some, and only some, principal factors are identified, it is called partially true (or, symmetrically, partially false). A statement is said to be principally false if it presupposes the image of the essential structure of a phenomenon described by it which contains some secondary factors for that phenomenon. At last, it is absolutely false if none of the factors it re-fers to is essential for the described phenomenon [more detailed

definitions, and comparisons with the other approaches to the approximate truth, are given in Nowakowa 1976, 1977]. Absolute truth reconstructs exactly the essential structure of a given phenomenon whereas the absolute false does not present any element of it. Among those extreme cases there appear those of relative truth, partial truth/falsity, and principal falsity.

One and the same general statement can change its essential-truth-value accordingly to the objective changes in the essential structure of a described phenomenon. For instance, Marx's law of value establishing a dependence between prices and values (amounts of socially indispensable work for producing given commodities) is, maybe, a relative truth for prices in free-market capitalism. But is turns out to be a principal falsity for prices in socialism where it is the state interest (of a rather non-economic nature) that decides about. And the value (in Marx's sense) plays at most an adventitious role for setting prices by the special state organ.

In order to put it a little more clearly [a closer analysis is contained in 1977a, p. 256 ff] let us consider the following idealizational statement:

(T) (x) (t) if $[U(x,t)$ & $p(x,t) = 0,$
            then $F(x,t) = f(H(x,t))]$

which presupposes the following simple image of essential structure composed of one principal factor (H) and one secondary (p):

H
H, p

as eternally valid — for every time t. Even, if in time t_0 statement (T) is an absolute truth (i.e., the hierarchy of factors influencing magnitude F is composed of principal factor H influencing F in f-like way and of secondary factor p alone), then it must not be for ever. For if in time t_1 a qualitative change in the essential structure of F appeared and factor q became a second principal factor for F, thus the essential structure of F took, say, the form:

H
H, p
$H, p, q,$

the thesis (T) would change its essential-truth-value turning out to be a partial truth. And if a further essential change appeared and factor H came down to the role of a secondary one for F, thus the essential structure of F took, say, the shape:

G,
G, H
G, H, p
F, H, p, q

then (T) would attain the level of the principal falsity.

It is obvious that in case of the first, quantitative, transformation the following transformation of (T) would remain absolutely true:

(T/t_0) (x) (t) if $((t$ is earlier than $t_0)$ & $[U(x, t)$ & $p(x, t) = 0]$,
       then $F(x, t) = f(H, t))])$

but such a temporal limitation of thesis (T) is not logically equivalent to the thesis itself. The latter, instead, changes its essential-truth-value along the categorial changes in the deep structure of the phenomenon it refers to. In that meaning an idea of the dialectical philosophy that "the truth is a historical category" seems not to be senseless.

3. The problem of the present paper may be formulated thus: to consider an epistemological evolution of Marxism in socialism, that is, to find out a correspondence between changes in the deep structure of a socialist society and variations of the essential-truth-value of Marxism understood as has been above explained.

Needless to say, such a formulation of the task of the present paper requires to adopt assumptions which need not be true. In order to compare the Marxist image of socialism with the reality of that system, one must have a given theory of the latter at his disposal presupposing it as being at least relatively true. Moreover, a qualification itself is being proceeded under the so strong idealizing conditions [Nowakowa 1977, p. 77ff] that a literal applicability of those more precise definitions is limited to rather simple schemes (*ibid.*). If so, then the evaluation of the cognitive development of Marxism can be a mere relative and approximate. It does not imply that the task is entirely hopeless and no criteria confirming, or disconfirming, an evaluation of that kind exist. There is such a criterion: if some well-known empirical trends in the development of Marxism in a socialist society become more understandable provided such an evaluation, that is, if they cease to be "irrational" or "accidental", then the evaluation turns out to have some empirical support. Despite its relativity and despite its approximative character.

III.

1. As announced above, the proposal presented in this paper is dependent on a certain philosophy of history. Some of its main ideas are [cf. 1983]:

There are three conceptually independent class divisions: between the owners and the direct producers, the rulers and the ruled, and the priests and the indoctrinated. The former is roughly of the Marxian origin, therefore let us briefly explain the remaining two. Political rule consists in the fact that a certain minority is materially stronger than the remaining majority due to its control of the means of coercion. It is in the interest of the rulers to enlarge their control over the activities of the citizens, whereas the interest of the latter consists in enlarging the sphere of social autonomy. Roughly the same can be said of the third sphere of social life: monopoly control of the means of indoctrination is the origin of the class division between priests and the indoctrinated inasmuch as the former (religious priests as well as "priests of mass culture") impose with the aid of these means their worldview on the latter.

Class societies are those in which the three class divisions are separated from one another. That is why the proper model of slave, feudal or capitalist society is not the Marxian economic two-class society but one with three classes: owners, rulers, priests, dominating the people on different planes.

Supra-class societies are societies in which class divisions become cumulative. Thus in a totalitarian order those who control the means of coercion have at the same time monopoly over the means of production, constituting thereby a dual class of rulers-owners (this is the actual sense of the expression "nationalization of the economy"). However, there still remains the class of priests. In fascism the masters of the means of coercion acquire monopoly control of the means of indoctrination constituting thereby the dual class of rulers-priests. However, the class of owners remains. Finally, if those who control the means of coercion simultaneously monopolize the means of production and indoctrination, they constitute the tripartite class of rulers-owners-priests dominating the people on all planes. Let us call such a society a socialist one since this is the name it gives to itself.

2. Let us inspect more closely the nature of the division on the political classes [cf. 1987]. The range of influence of a single ruler will be termed the set of citizens' acts controlled by the ruler. The sum of all the ranges of influence of particular rulers forms the range of regulation of the class of rulers as a whole. Competition between the members of that class forces every ruler to enlarge his or her sphere of influence — those who falter lose their position and cease to be rulers. The common interest of the class in question consists in fact in widening the scope of regulation, and as such is at variance with that of the class of citizens. All the more so in that an increase in the scope of regulation necessitates increased repression.

We will term civil alienation the ratio of the number of acts from the sphere of regulation to the number of actions undertaken by the citizens. It is presupposed that the level of civil resistance — conceived of as the percentage of citizens who break the relation of subordination to the rulers — is determined basically by civil alienation. When civil alienation is low, the level of civil resistance is also obviously low. A certain regulation of social affairs, even if secured by the use of physical force, is perceived as necessary and a significant majority of citizens do not oppose it. But contrary to what may appear to be the case, the level of resistance is also low under very high levels of civil alienation. Extensive regulation followed by a high degree of repression leads to the collapse of social bonds which precludes any possibility of common action. There are thus two ways of achieving the social peace. One is to retain the balance between the citizens' aspirations to have a large amount of autonomy and the regulation of social life on the part of those in power (the state of class peace). Second is to control citizens sufficiently to make any kind of resistance socially impossible (the state of declassing). The state of declassing contains as its sub-interval the state of totalization, under which the scope of regulation covers the entire field of potential control of power over the citizens.

The level of civil resistance is highest if civil alientation is "at a medium level", that is to say, painful enough to evoke social reaction but not to paralyze civil society. This may be termed the revolutionary level of resistance but one should keep in mind that such revolutions may take different forms, both violent and peaceful.

In sum, the dependence in question can be represented in the form of a bell-shaped curve with the state of class peace on the left side, the central revolutionary area and the state of declassing on the right side. This dependence presents a static image of the relationships between the two political classes. However, over time, it is assumed, that the image changes. Initially, terror crushes social bonds and autonomous social relations of the type "citizen-citizen" are replaced by those of the "citizen-power-citizen" type, thereby making any kind of resistance impossible. Yet autonomous social relations gradually recover and civil society gradually comes back into being. Accordingly, civil alienation diminishes since citizens undertake ever more actions outside the forms dictated by the rulers. As a result the social ability to resist gradually increases and finally reaches the revolutionary interval as more and more categories of citizens extricate themselves from the state of declassing. The process may be termed the revalorization of social bonds, or the recovery of civil society [cf. 1987].

3. The starting point for the proposed model of socialist society is the assumption that socialism sonsists in the triple monopoly over all types of material means of control of people's actions. It is of some importance to notice that not all of the material characteristics of the triple ruling class are equally significant as far as its manner of domination is concerned. For first and foremost they are rulers who also dispose of the means of production and indoctrination. That is why, members of the triple-ruling class can be modelled as rulers and the system of triple rule as a purely political society. In other words, an idealizational theory of a socialist society is a sequence of models of increasing realism with the purely materialist model of power as the initial one.

Thus we claim that the society under consideration is formed of only two political classes, viz. the class of those who dispose of the means of coercion and the class of citizens. The two remaining class divisions, not to mention secondary social categories (e. g., social strata), are neglected in our model. The sphere of politics is then treated in total isolation from those of the economy and culture. Moreover, the influence of political institutions and society's political consciousness on political class relations are put out of play.

Of special importance is the assumption that the considered society is totally isolated from remaining societies. There are also additional idealizing conditions in force. Thus it is assumed that there is no technological progress in our abstract society, that only simple reproduction of the population occurs, etc. In general, all factors not referred to in the following summary of our model are neglected.

4. Consider how a single purely political society develops in time. Because of obvious reasons, only theses will be put forward, and only these which will be relevant for the task of this paper. A fuller justification of them may be found in [1987].

(1) Let us suppose that in the initial stage of development of our abstract society a state of class peace reigns. This implies a low level of civil resistance. Yet competition between those in power forces the typical ruler to enlarge his own sphere of influence. Otherwise he will be wiped out in the competition. As a result, the range of regulation increases and hence civil alienation also grows and, after a time, reaches the revolutionary interval.

(2) The revolution may be defeated, or possibly the citizens may win. In the first case post-revolutionary terror appears. Its social function, independent of the intentions of particular rulers, is to atomize the citizens. In this way the only factor able to stop the pressure of power, namely civil resistance, disappears. In the second case, nothing at all changes significantly. For the revolutionaries constitute simply a power elite inasmuch as they dispose of a new means of coercion; for the masses in revolt, not to mention the armed civil guards, are nothing but a means of control. Thus after the victory, the revolutionary elite seizes monopoly of the means of coercion and becomes a new class of rulers. Appropriately it undergoes the same regularities as the previous regime, including the basic one: the trend towards increasing their power. For among new rulers there always appear people who crave power for power's sake, and they will win in the competition with the "passionate revolutionaries". As a result the same incessant increase of the range of regulation appears as it did under the old regime. This in effect institutes a new wave of civil resistance directed against the new class of rulers. In this way the civil loop is completed

which makes way for the same two possibilities: either the defeat of the civil masses and thus the postrevolutionary terror, or their victory and thus the second civil loop. Etc. The outcome is that after a number of civil loops the citizens lose definitively, and some repertory of the class of rulers carries out the declassing.

(3) The declassing of citizens enables the class of rulers to enlarge their regulation without any constraints so that all that can possibly be controlled becomes in fact controlled: the system reaches the state of totalization. In such a state there are no more areas of social life to subordinate. But a typical ruler must, as always, increase his influence. Under our present assumptions there is only one way to keep the whole system in operation — a purge. Purges eliminate some rulers thus leaving room for competition among those remaining and resulting thereby in a new expansion of political regulation. After a time, however, the expansion again reaches the state of totalization, and yet another purge becomes functionally necessary.

(4) It is the increasing ability of citizens to resist the system of oppression (on the strength of the adopted model of revolution) which breaks the vicious circle of purges. What appears is the first, as yet limited, revolution embracing that part of the population in which the process of revalorization of social bonds was the earliest. It gives the rulers an opportunity to extricate themselves from the periodic purges and introduce a new system of relations between rulers and citizens based on the reduced sphere of regulation. Yet, the mechanism of competition works as before and the sphere of regulation inevitably grows. If it does not reach the state of totalization again, it is due to the growing ability of citizens to resist (on the strength of the adopted model of revolution). As the counteraction against the again increasing regulation, a new revolution comes about. It is more of a massed revolution than the first for new categories of citizens have extricated themselves from the declassed state. It will again lose when the rulers reduce the sphere of regulation in order to avoid yet another revolution. Because of its greater social base, the second revolution is also marked by an increased chance of victory, that is, by a civil loop. This implies that the society repeats its historical road from the very beginning, including the declassing and totalizational stages. Yet, under our idealizing conditions the chances of victory for the citizens are not very high and the standard image of the movement of the society under consideration is the following: revolution — declassing — increasing regulation — revolution on a wider social basis, etc.

(5) Thanks to the recovery of civil society, more and more citizens extricate themselves from their declassing state. There comes a point when the large majority of citizens is ready to oppose the system, such that it proves socially impossible to implement the usual declassing strategy. If the civil loop is again avoided, the only possibility for the rulers is to make concessions. The range of their regulation diminishes rapidly and the sphere of social autonomy increases correspondingly. Let us term this the revolutionary inflexion, for it brings the system from the state of declassing to that of class peace.

But the mechanism of competition forces the class of rulers to recover the lost social territories, which results in the next revolution, etc. In this way a new pattern of development comes into existence: revolution — concessions — revolutions on a wider social basis — increasing concessions, etc. This cycle of the reappearing class peace continues until the state of social peace is permanently attained, that is, until a new system of power gradually evolves that will limit regulation to a minimum, and thus be acceptable significant that will be limit the regulation until the minimum and thus will be acceptable to the citizenry.

5. The foregoing model is highly abstract, yet may help to explain certain historical trends in the Soviet Union as far as the crucial dimension of it, the social relations between the rulers and the civil masses are concerned.

The initial point is the system close to totalitarianism [cf. 1983, part. II] which met resistance on the part of the masses in the February revolution. However, the new, mainly social democratic, policies combined to a large extent disposal over the means of coercion and the means of production and, despite their democratic ideology, enforced this fusion. On the eve of mass resistance the Bolsheviks seized rule and in this way completed the process of the cummulation of power and property. The October Revolution was the second civil loop. As such it led to the subsequent revolution, namely to the workers and peasant uprisings in 1921. Declassing comes as late as the end of the twenties. The masses are pacified as a result of mass terror and collectivization. A system of social terror emerges. The social basis of such a system was, then, the Gulag. Yet steadily, independent social relations actually begin to recover and a wave of strikes and finally armed upheavals appear there at the beginning of 50ties.

The system had to be changed if it was not to be destroyed in its very foundations. The Gulag was pacified by liberating political prisoners and allowing them to merge in the still captive society. From then on the Soviet Union has been, if we may conjecture, in a phase of cyclic declassing. The next wave of resistance appeared at the turn of the sixties: the army was forced to intervene in 14 towns of the Soviet Union. At the end of the seventies and at the turn of the eighties a wave of strikes occurred.

Let us notice that the social history of some other socialist countries also falls under the outlined image, although they are at different stages of the process [cf. a larger analysis in 1984].

IV.

Having assumed all that, one may try to evaluate changes the contents of Marxism (understood as defined above) have undergone along the development of an ideal socialist society (as presented above in the summary of our simplified model). I would like to repeat that the assessment in question is a relative one, that is, it is in force on the condition that the adopted stand, viz. non-Marxian historical materialism, is acceptable. If it is not, the below analyses are mistaken. But I take a risk to admit it is.

1. Russia of the second half of 19th century is a scenery of a rapid stateisation of her economy — feudalism is being transformed into the state capitalism without any intermediary stage of free-market economy. As a result, the totalitarian tendencies that occur in Western countries as late as in our time took place in Russia at the beginning of the century [see 1983, Chaps. 18-20]. Therefore, Russian society that never could be grasped sufficiently with the aid of Marxist categories, breaks them entirely. Marxism applied to the undergoing totalitarisation of society brings from the very beginning quite erroneous results. For it inclines to treat the dominating for centuries Russia's social force, viz. the state, as a representative of feudals alone, whereas the tsarist state was simply representing its own interest, that is, maximisation of power over its, and the foreign if possible, citizens. Such an interest of a material — for resulting of the monopoly for forces of coercion — but of non-economic nature is simply incomprehensibe in Marxist

categories. Applied to specific social conditions of Russia, who for some times in its history (Ivan the Sinister, Peter the Great) was revealing totalitarian tendencies, Marxism with its means adopted to the class society alone was simply cognitively powerless. That is, it was from the very beginning a principal falsity. Nothing surprising that actually Marxism may be found on the banners of those who proceed Russia, at last effectively, to totalitarianism. It made it possible to present a totalitarianism and then socialism as a march toward the classless society.

2. Such was, particularly, the role of Leninism. Not because of being a "deviation of the true (young) Marx's thinking" but quite the contrary so, because of being quite faithful to the (mature) Marx's ideas, first and foremost to the Marxist economism.

And so, nothing but a right application of Marxism to Russia's conditions was Lenin's thesis about "backward tsarist Russia" put forward just at the time when Russia, in the light of actual historical processes, was already ahead of the historical development distancing the Western countries as far as the cummulation of class divisions is concerned. Nothing but a natural application of Marxism was Lenin's interpretation of February Revolution as a bourgeois one, whereas the latter was simply last chain of the rising double-class of rulers-owners towards the exclusive power. It was also quite admissible in the light of teachings of Carl Marx to interpret the October Revolution as a (truly) socialist one, whereas it was a civic loop changing the personal composition of the double class of rulers-owners alone and enabling the new garnitur of it to seize the means of production of consciousness. The latter began to form social consciousness in the spirit of Lenin's ideas that were already then principal falsifies.

First and foremost, however, the historical merit of Leninism for the system of triple-rule is that it allows not to understand the nature of revolutionary process. For as long as one accepts Marxian historical materialism and Leninism based on it, the program of building a socialist society seems to be quite a rational: the working class fighting against the capitalist oppression calls into being its avant-guard, the communist party, which seizes the political power and on the strength of new apparatus of the workers state annihilates the private property replacing it with the socialist relations of production. On the ground of the non-Marxian historical materialism things look quite differently so: of the working class a political elite emerges which seizing the monopoly for the forces of coercion transforms into the class of rulers undergoing the same objective mechanism as the one that has been actually overthrown. And even the mechanism of maximization of power is reinforced within new, revolutionary garniture of rulers because they have obliterated the most powerful controller of the political authority, viz., the private property. That is why, the standard mechanism of political competition forces them to take, independently of their political program, the disposal means of production over, that is, to become a double class of rulers-owners. And later on, the triple class. Leninism was the best candidate for the ideology of the growing system actually because it did not understand anything from that conceptualising the process in question in quite erroneous terms. The point lies thus in the fact that Leninism was principally false, not in Lenin's intentions. The latter were perhaps not too bad, as his condemnation of what he was able to perceive a mere as "bureaucratization of the Party" testify. That is why, the interest of the triple-ruling class required to impose that Marxist lack of understanding of what was taking place in Russia to the people masses. In short, Marxism has not realized the fact that every victorious revolution is a civil loop.

3. The system of triple-rule achieves its maturity when the masses are already declassed and no autonomous social bonds exist, that is, when all the relationships of the man-man type are being transformed into relations man-authorities-man. It was a time of the so-called "stalinism" when the political power was intervening, as it is known, even in the relationships between parents and children. That was, then, the mature socialism. The mature socialism is that of the desocializing phase.

And it is Stalinism which corresponds to that historical period on the ideological scene. Stalinism was in principle as faithful to the Marxian original as Leninism was. To be true, it was cognitively much more primitive and scholastic. But it scholastically petrified the original Marxism. And actually thanks to this — and not to less or more alleged "deformations of Marxism" — Stalinism was able to play its ideological role very well. It was in full agreement with the interest of the power to tell the masses of the country full of camps of slave work that they live in the most free society in the world. Who had revealed

a slightest doubt identified himself as an independent citizen dangerous for the power. And, in fact, he was. For that one who has enough
courage to reveal his doubt in an open absurd in the conditions of
immensurable terror, will become a revolutionary when the terror
diminishes a little. Stalinism was a means to maximize the distance
between reality and the Marxian ideal. One who dared to notice that
distance identified himself as an opponent of the system.

4. But the protest of most oppressed masses leads to the first, still
local, revolution and takes the system of the phase of declassation.
The phase of cyclical declassations begins interrupted with revolutions
of an increasing social basis. The system starts to develop in the
rhythm of the more and more intensified class struggle of the people
against the triple-lords. And the old Marxist categories of the
exploitation, the class division, the class struggle, socialisation of
means of production etc., become recovered. The social world of the
system of triple-rule begins to reveal processes that are described in
the Marxist model of social development: strikes and upheavals in
Soviet concentration camps at the treshold of the 50-ties, the general
strike in DDR in 1953, the strike in Pilzn, Czekoslovakia in the same
year, workers revolt in Poznan, Poland, in June 1956, the Hungarian
revolution of October-December 1956. The phenomena could be easily
interpreted as the class struggle of the masses against the new class of
owners, which implies the applicability of Marxism to new social
conditions of the so-called "socialism".

And, in fact, there appeared a conception [Djilas 1958] that was
explaining the genesis of socialism in Marxist terms. According to
Djilas, underdeveloped countries, in particular Russia, were facing the
following alternative: either to industrialize themselves or to resign of
the active role in history falling into the servitude of more economically
developed countries. Their home capital and classes and parties representing it were too weak to solve the problem of a rapid industrialization efficiently enough. In those countries it was a mere revolution that
was able to satisfy vivacious needs of a nation and that is why the revolution became there unavoidable. The only social force which could
make it was the proletariat, that is, the revolutionary party representing
the working class [Djilas 1958, pp. 18-19].

Socialism is, then, a means to reduce the long lasting backwardness
resulting from the delay in industrialization of eastern countries. It is

a purely Marxist explanation. And so is Djilas' analysis of the social structure of the new system. For it is the appearance of the "new class of owners" which is a price for the rapid industrialization the socialist countries are forced to pay. The class is formed of the party functionaries but its constitutive feature is the collective ownership of means of production; the political power and the outlook monopoly are the means alone to keep the new class of owners in their economically privileged position.

Socialism turns out to be then, according to Djilas' analyses, a new form of a class society based on the economic inequality in the same manner the older class societies were founded upon. If so, the only social force — it is an equally Marxist conclusion as its premisses — able to overthrown the new class of owners is the proletariat. Actually in socialism Marx's idea of the proletarian revolution becomes vivid anew.

As it is seen, this purely Marxist standpoint agrees with empirical facts of the late socialism quite well so. In the late socialism the essential-truth-value of Marxism grows significantly. For the economic position of an owner is only one of three faces of a triple-lord and it is not so that the political power is a means only for the increase of the economic exploitation. Quite the reverso so, it is the economic exploitation which is a means only for the enlargement of domination of triple-lords over the masses. Nonetheless, the Marxist conception gains in the late socialism a dignity of a serious candidate for the explanation of the social nature of the system: Marxism becomes partially true. From the principal falsity to the partial truth/falsity — such is the categorial line of the development of Marxism in the socialist society. It is a line of the epistemological progress.

And that is why Marxism ceased to be a suitable means to support the triple-ruling class interests. These interests were even requiring significant changes in the contents of the ideology of the system that was cognitively recovering. Marxism became a danger since it turned out to contain too much a truth about the late socialism.

5. It is a ground that makes, I guess, more comprehensible the social role of the wave of the socialist revisionism that appeared in several socialist countries after the "stalinist period". For what was the worst on the ideological scene for the system of triple-rule in the phase of cyclic declassations was keeping Marxism in the role of the official

doctrine. At present, when the class struggle Marxism is recalling to is being transformed into a normal phenomenon of the socialist society, the category of a class (even if understood only economically), of a revolution (even against the economic exploitation alone), the moral pattern of a revolutionary — all that could begin to operate against the system of triple-rule.

Fortunately for triple-ruling class, the moral protest of the leftist intelligentsia against Stalinism took a purely idealistic form: it has been directed not against the material interest of triple-lords but against the Marxist ideology with which the system covered itself. Against the fundamentals of Marxism (understood as it is adopted here), those intellectuals of the Marxist origin have not identified the ideology of Stalinism as a cloud the system stretches around to hide its genuine nature but as a self-contained cause of "deviations". And the notion of "deviation" presupposes a pattern from which reality is departing. In that case the pattern was not discovered in reality but deducted from Marx's own writings.

The latter were, however, too close to Stalin's one. The historical accident has helped the triple-ruling class, namely, the discovery of early Marx's writings Marx himself had not recognized for being worth published for forty years of his mature scientific activity. After the years of absence in the socialist philosophical literature Marx's *Manuscripts* of 1844 were introducing into circulation. And this gave an opportunity to reinterpret Marxism entrirely, that is, to change it into something else[2].

First and foremost Marxist theory of history has been rejected. The category of a class was replaced with that of a "species of human being". And instead of Marxian laws attempting to give general conditions of a transformation of one socio-economic formation into another, the young Marxian philosophy of history with its idea of communism as a movement abolishing the divergence between human "existence" and "essence" was introduced. The divergence between the two, that is, the alienation turned out to be the motor-force of history, not the protest of masses against exploitation. But everybody may be alienated, including triple-lords themselves. They are also, even quite often, in a position of a "pupil of the magician" who liberate forces which they do not control. The category of alienation being applied on both sides of the division into triple-rulers and the people was concealing the social nature of socialism. And the triple-ruling class actually needed something of the sort. The replacement of an ideal of revolutionary with vague thoughts about "self-realisation" of a human being was only a consequence of the historiosophic revision of Marxism.

The new doctrine had rather little to do with Marxism, but was the best of then existing alternatives to deform Marxism into direction necessary for the weakened triple-ruling class. The revisionists protesting against immoralities of the stalinist period were unconsciously saving the official ideology of the system of triple-rule.

6. Needless to say, that was a social process having place in its decisive parts and aspects outside the consciousness of its participants; as usually social processes do. It is almost unlikely that any of members of the triple-ruling class understood the matter and so did not the revisionists. The ruling classes, however, not being theoretically illuminated, behave in a practically wise manner. It was quite sufficient for those in power in Poland, Yugoslavia, and later in Chekoslavakia and Hungary, to feel that after so many revolutions of the beginning and the middle of 50-ties the mature Marx's ideas about the class struggle that is always justified morally and leads the progress in history, i.e. the ideas that were treated for so long a time as empty watchwords alone, become danger. The more so, that they are their own, the authorities's, ideas. In that situation every reinterpretation of Marxism that ceased to assign it those danger contents was welcome. Since a humanistic reinterpretation of Marxism appeared, it was welcome as well. The more important thing was to eliminate from Marxism its revolutionary vision of history and its axiology of a revolutionary struggle. The strongest moral condemnations of Stalinism were much convenient for the triple-ruling class than an interpretation of strikes and upheavals in Gulag in terms of the class struggle and the party apparatus in terms of the ruling class.

However, the followers of the anthropological interpretation of Marxism (Warsaw school of historians of philosophy, Budapest school, "Praxis" group in Yugoslavia) were too consequent thinkers and deformed Marxism too much. And no ideology can allow for an entire break with the history of its own. Also their consequence was inadmissible to a state ideology which had to be eclectic in order to give rationalization of a policy for possibly different, and even unpredictable, occasions. That is why, all the revisionists underwent serious repressions after they did what they had to do in the interests of system they had not understood at all.

An objective, unintended result of the intellectual activity of revisionists was a new form of the official ideology of the triple-ruling class. For their influence upon Marxists in socialist countries was so great that the latter were changing their minds in the direction shown by revisionists, that is, in a way advantageous for triple-lords. As a result, Marxism was transformed into the "socialist humanism", a vague and unclear doctrine that is an eclectic combination of those elements of Marxian historical materialism that are not danger for the interests of triple-rulers (e.g., the conception of productive forces and relations of production, economic base and political superstructure, etc.) and of the moral doctrine of young Marx. The doctrine is not Marxist — under the adopted understanding of the term — at all. It is a conception resulting of the Marxist heritage deformed in the manner shown by revisionists who unintentedly gave the triple-ruling class the means to take the masses away from spiritual inspirations arising from the Marxist tradition.

The people class of socialism obtained, however, from the "Marxist socialist humanism" a new wave of indoctrination. In fact, in contradistinction to Marxism, the new eclectic doctrine was able to play the role of ideology being again on the level of principal falsity.

7. In recent times a new intellectual formation in socialism has signallised its existence, namely followers of the idea of the "socialist democracy". They were active in Poland during the 1980-81 revolution as proponents, on the political platform, the so-called "horizontal structures" within the communist party. Their leading idea was that Marxism ascribed too great a significance to material (identified by them with economic ones) factors, whereas the course of the newest history has revealed that the role played by social institutions is not negligible at all. In particular, if it was possible in Russia to install from the very beginning the form of democracy instead of dictatorship, the course of socialism, and the influence of socialism in the world, would be of a quite different nature. However, because of the historical heritage of Russia, and the influence of the Soviet Union upon the formation of new socialist systems, it was impossible. But today, after the "great economic industrialization" associated, unfortunately, with the rejection of any democratic forms of

organization, there is no economic need to keep the out-of-dated system of the "dictatorship of the proletariat". Quite the contrary so, actually that system turned out to be an obstacle to further socialist development; the requirement of democracy reveals to be a necessary condition of further economic development of socialism. Democratic institutionalization of the political life turned out, somewhat surprisingly as seen from the orthodox Marxist standpoint, to be more fundamental than the development of technology which does not imply automatically most serious problems the "realised socialism" faces.

This trend of social thinking appears at a time when the mass resistance of the people against the triple-ruling class becomes the most decisive factor of social life. Then, the interest of the system requires not to allow the masses to unit themselves on the basis of an antagonist vision of society they live in, that is, not to allow to an appearance of class consciousness among the masses. For such a consciousness would lead them to crushing the material foundations of the system of triple-rule. Every other direction of the revolutionary pressure is better than that. Paradoxically, it is also better for the civil masses themselves. For, if the conception of civil loop is true, the victorious revolution would be, indeed, a cataclysm for the civil masses, leading merely to replacement of triple-rulers of one ideological faith with triple-ruling class of another faith.

The communist democrats were thus counteracting, consciously or not, the victory of the masses (i.e. the civil loop) when spreading the institutional grasp of society. In this way a quite strange to Marxism — but peculiar to Liberalism — institutionalism is added to the body of official ideology. The All-national State devoid of any class contents and, what follows, class enemies is put forward as a central institution of social life which should be improved, step by step, with legal measures but cannot be overthrown. For at the same time it is a central value of the "socialist morality".

In that way Marxian historiosophy of the class struggle based on the materialist "dialectic of productive forces and relations of production" was at first replaced with the idea of abolition of alienation and next with the institutionalist grasp of society. In the first step the class dimension of Marxism had been rejected, in the second one the same has happened with the materialist grasp of society. For both were indispensable to keep the official ideology of socialism on the level of principal falsity as the interests of ruling classes require.

V.

If our assumptions are true then the described case implies some conclusions worth perhaps to be noted.

1. First of all, the notion of ideology we were using quite loosely above, seems to become a little more clear. For a social doctrine a necessary condition to be an ideology of a social class is to be principally false. In other words, an ideology is to give a mistaken image of the essence of phenomena it is about, but the image is to possess the "appearance of truth", that is, some features of the phenomena in question are to be recognized property. For the social function of ideology is to justify a privileged position of the social class it serves to[3]. Now, if our initial assumptions are correct, then the ruling classes (of owners, of rulers, of priests or their multiplications the triple-class included) possess an actual interest in the lack of knowledge concerning the hidden nature of social life. Members of those classes cannot be aware of being exploiters and/or oppressors operating on social or spiritual life of subordinated people. Otherwise they would play their role of being them less effectively than they do and even they could leave the ranks of their own class as it happens to some future revolutionaries. The more so this concerns members of subordinated classes: the stability of every class, or supra-class, system requires of them to be unaware of being subordinated at all. That is why, as for the hidden nature of social life (the class, and supra-class, divisions and conflicts) it is a functional reason of the existing social system to impose all its members, including privileged ones, the "false consciousness", that is, principally false social doctrine, that is, the one which distorts entirely the hidden nature of society but at the same time says truth about adventitious elements or aspects of it.

A necessary condition for a doctrine to remain to be an ideology is to continue to be principally false. And if reality the ideology is about is changing in such away that its contents become more and more essentially true, then a defensive reaction of the ruling class is to change appropriately the ideology of its own, so that the level of the principal falsity is kept. We have seen this in the history of Marxism in socialist society.

2. All the existing societies are either class or supra-class ones and hence their social structure is basically polar. And so is the structure of their social consciousness[4]. It is being composed of ideology dominating in

normal conditions over the majority of society but at the same time always occur elements of anti-system social thought becoming spread and even dominated in revolutionary periods. And as ruling classes have an interest in hiding the nature of their systems for it is injurious for the people masses as the masses themselves because of the same reason have an interest in revealing that. In other words, what is a deep structure of a (class or supra-class) society is to be kept in the shadow for ruling classes and is to be laid open according to the interest of oppressed classes.

Yet, the revolutionary utopia has also something to hide. It namely conceals the social mechanism of the victorious revolution which turns out inevitably, if our assumptions are true, to be a civil loop. Utopia hides then what is disadvantageous for the revolutionary power.

A necessary condition for a doctrine to be an utopia of the oppressed masses is therefore to be partially true/false and hence to reveal the actual image of the class division but not to understand a role the victorious protest of the oppressed masses is playing in social life, i.e. to conceal the civil loops.

It is obvious that the epistemological conditions in question for both an ideology and utopia may be satisfied by many doctrines. Therefore, a sufficient condition of being an ideology (resp. an utopia) is perhaps the one recalling to social conditions of spreading of a given doctrine. Maybe of different doctrines fulfilling the epistemological condition, that is, being principally false, that one becomes the ideology of the ruling class in a given time that possesses the greatest persuasive power, i.e. the ability of influencing the largest amount of people in a given time. But the matters cannot be discussed here so carefully as they deserre to be. Let us rather formulate some conclusions that follow from the proposed (necessary conditions for) notions of ideology and utopia on the ground of the presented grasp[5].

3. The above characteristic of ideology and utopia is adjusted to the requirements of non-Marxian historical materialism. It allows, however, for a certain generalization. And so, not only ideology (of the oppressive classes) but also utopia (of the oppressed classes) are to be classified as — let us take this term — ideolusions. Ideolusions possess not only classes and individuals but also nations, civilizations, races etc. Also mankind as a whole possesses ideolusions. An opinion that "man is a child of God" is normally welcome whereas one that man maximizes domination over neighbours is rejected. In both cases, however, this has little to do with

the substantive reasons and the only motive for acceptance (resp. rejection) of an idea is whether we, people, feel satisfied with the self-image which they offer us.

A general notion of an ideolusion covering all these cases presupposes the general notion of interest[6]. Having somewhat counterfactually assumed the latter is done, the construction would proceed as follows. Given is property F and its essential structure S/F. An upturned image of the structure S/F will be termed such an image of S/F which as the main properties for F claims to be those which, in fact, are at most secondary for it. An ideolusion of (a person, a group, etc.) X in regard to property F is such an idea of F which presupposes an upturned image of the essential structure of F and, moreover, actually due to this the idea under consideration rationalizes some X's interests.

4. The shortest history of Marxism can be put forward as follows. Within capitalist society Marxism is a revolutionary utopia but transforms into a new ideology along with the transformation of the revolutionary elite in triple-ruling class. But on the strength of mechanisms of the development of socialist society Marxism is gaining more and more elements of the people utopia, that is why it is to be distorted and mixed with other doctrines, so that this eclectic composition retains the level of the principal falsity, that is, the ideological level. That this is done with hands of the less or more overt opponents of the system of triple-rule belongs to the domain of the "irony of the historical process".

Dept. of Philosophy
A. Mickiewicz University, Poznań,
ul. Szamarzewskiego 89, Poland

NOTES

[1] The epistemologically oriented idea of sociology of knowledge as applied in the present paper follows to a large extent an approach developed by Buczkowski [1980], [1987].

[2] As many other assessments of the kind, this follows from the fact that we have decided to use the term "Marxism" to denote the theses (c) – (ccc) throughout the whole paper. Clearly, under another definition of Marxist social philosophy, the conclusions of what is Marxism, and what is not, could sound differently. For instance, if one adopts the interpretation of Marxism presented by Lukács [1923], then the theses put forward above would be unsound. This is, however, a normal thing in sociology of knowledge that its theses depend both on the image of the social reality which is hypothetically adopted and on the hypothetically accepted interpretation of the social doctrine under consideration.

3 According to the proposed grasp it is the way of idealizing the social phenomena which is decisive for a conception to be ideological (or utopian). The linguistic form of "ideological statements" is instead of no importance at all. One may conjecture that the form in which ideology (and utopia) is expressed changes historically – it was the religious language in the Middle Ages, whereas in our times ideologies are formulated in the language of science. Therefore, including into the definition of ideology any explicit reference to the language it employs implies merely a superfluous constraint on generality of considerations [cf. Hochfeld's 1963 criticism of this kind of approach in the theory of social consciousness]. Sometimes such reference is too rigorous in itself, e.g. when it is postulated that ideological are such statements that imitate scientific elements while containing nonscientific elements [Geiger 1953, p. 66]. Apart from the limitation shown above, such definition is inadequate because of the fact that very often quite legitimate statements of social sciences simultaneously possess a clearly evaluative character. For instance, the economic theory of free market describes not only the ideal type of a certain kind but the liberal (positive) social ideal as well; the theory of surplus-value in Marxism describes not only the ideal type of capitalism but the Marxist (negative) ideal at the same time; etc. The point is not to eliminate these theories from science as "ideological" but to observe the standard scientific rules in testing them against empirical evidence [cf. more 1980, p. 218ff]. This is the real sense of M. Weber's [e.g., 1949, p. 95] position as to the status of value-judgements in science: they are admissible, but in science they must be justified cognitively and not morally [cf. Kmita 1976].

4 Even when remaining on the grounds of Marxian historical materialism one might see reasons for making distinction between (a) worldlook, (b) science and (c) class consciousness and treating them as irreducible natural kinds in the field usually labelled "social consciousness". This is seen, I believe, if one gives a definite interpretation to Marxian idea that "social being determines consciousness". As I have already had occasion to argue [1975, p. 89-91] this idea can be reconstructed as the following formula:

(F) out of historically given systems of ideas, this one becomes widespread in a given society which, for a given state of the socio-economic conditions, ensures the greatest stability to the optimum political system and the optimum system of relations of production; the "optimum political system" being the one maximizing the effectiveness of introducing the "optimum system of production relations" and the latter being in turn the one maximizing the surplus product under the given state of productive forces [cf. enlargements in Buczkowski 1981, 1982 or Łastowski 1978, 1982]. Now, formula (F) evidently fails for (b) as science does not adopt to the level of productive forces. Quite the contrary, it forms it [Cohen 1989, p. 4ff]. Whether formula (F) works for worldlooks is rather doubtful. Instead, a proper field of application of the formula (F) is that of class consciousness (c) and this field, following Bukharin [1923] and Benjamin [1934], can even be enlarged [cf. 1983, chap. 10]. Let us add that some Marxist authors interpret Marxian "determination" of consciousness by the "social being" as the relation of structural analogy between them [e.g., N. Egebak 1974, p. 256ff]. Yet, if "determination" means structural analogy, there is no actual reason why the economy is to be privileged as the frame of reference for analogies. Certainly, there exist other domains of the world which are isomorphic, e.g., with the structure of a given worldlook. The focus on the economy might be justified on the condition that it is presupposed in quite a different sense of "determination" than the one under consideration [e.g., causal or functional – cf. an analysis in Kmita 1971, chap. 3, 1975] and it is supposed that the source of the other "determination" is the economy. Yet, in this way Marxian historical materialism is made an ambiguous conception and "social consciousness" all the same remains an unhomegeneous sum of different sociological natural kinds.

5 Some authors apply so general notion of "ideology" that the difference between a wordlook and class consciousness disappears altogether [cf., e.g., Althusser 1970, p. 137, 150ff.]. However, even assuming Marxian historical materialism there are still reasons to differentiate between the two (cf. note 4). Some other authors preserving the conceptual difference between the two notions, claim that the factual difference between their designates gradually disappears as in modern times it is ideology which plays the role of a worldlook [e.g., Mannheim 1937, p. 36ff, Habermas 1973, p. 12]. The increasing influence of ideologies (and utopias) in modern times is a fact. It seems, however, that there is a significant difference between such a form of social thought whose subject is the social world itself and such which models an individual's life. Even if it is so that a vision of the natural cosmos is being replaced by a vision of the social world in our worldlook, the latter remains a "regulating principle" justifying existential choices which an individual is compelled to make in his/her life. And this still differs from justifying the social choices made by people as members of social communities. These two types of justification delimit the difference between a worldlook and an ideology, even if one and the same doctrine occurs in these two roles.

6 And not that of goal [like, e.g., in the definition given by Quinton [1973, p. 14] as this limits the applicability of the introduced category to, at best, individual acts.

LITERATURE QUOTED

Althusser, L. [1970]. Ideology and Ideological State Apparatuses (Notes towards an Investigation), trans. from French, in: L. Althusser, *Lenin and Philosophy and Other Essays*. London: NLB.

Benjamin, W. [1934]. "Der Autor als Produzent", after the Polish translation in: W. Benjamin, *Twórca jako wytwórca* (The Author as Producer). Poznań: WP.

Berger, P. L. [1966]. "Identity as a Problem in the Sociology of Knowledge". In J. E. Curtis, J. W. Petras, eds., *The Sociology of Knowledge*. New York/Washington: Praeger 1970.

Buczkowski, P. [1980]. O potrzebie krytycznej teorii socjalizmu (On the Need of the Critical Theory of Socialism). *Poznańskie Broszury Społeczne*, 2. Poznań: WiW.

--- [1981]. Z problematyki teorii społeczeństw ekonomicznych. Szczecin: PAM Press.

84

--- [1982]. Toward a Theory of Economic Society. An Attempt at the Adaptive Interpretation". In *Social Classes, Action and Historical Materialism (Poznań Studies in the Philosophy of the Sciences and the Humanities*, 6). Amsterdam: Rodopi.

--- [1987]. Uwagi o strukturze świadomości społecznej (Remarks on the Structure of Social Consciousness). *Kultura i Społeczeństwo*, XXXI, 4, 145-162.

Bukharin, N. [1923]. *Teorija istoriceskogo materializma* (The Theory of Historical Materialism), after the Polish translation. Warszawa: Roj 1928.

Cohen, G. A. [1989]. *History, Labour and Freedom. Themes from Marx*. Oxford: Clarendon Press.

Djilas M. [1958]. *Die neue Klasse*. Muenchen.

Egehak, N. [1974]. Ku materialistycznej teorii tekstu (Towards a Materialist Theory of a Text), trans. from Dunish. In A. Lam and B. Owczarek, *Marksizm i współczesne literaturoznawstwo* (Marxism and the Modern Science of Literature). Warszawa: PWN 1974.

Geiger, Th. [1953]. *Ideologie und Wahrheit. Eine soziologische Kritik des Denkens*. Stuttgart/Wien: Humboldt.

Habermas, J. [1973]. *Legitimation Crisis*, trans. from German. London: Heinemann 1976.

Hochfeld, J. [1963]. *Studia nad Marksowską teorią społeczeństwa* (Studies on Marxian Theory of Society). Warszawa: PWN.

Kmita, J. [1971]. *Z metodologicznych zagadnień interpretacji humanistycznej* (Methodological Problems of the Humanistic Interpretation). Warszawa: PWN.

--- [1975]. Marx's Way of Explanation of Social Processes, *Poznań Studies in the Philosophy of the Sciences and the Humanities*, 1, 1, 86-90.

--- [1976]. *Szkice z teorii poznania naukowego* (Essays from the Theory of Scientific Cognition). Warszawa: PWN.

Kołakowski, L. [1978]. *Main Currents in Marxism*. Oxford: Oxford Univ. Press.

Łastowski, K. [1981]. *Problem analogii teorii ewolucji i teorii formacji społeczno--ekonomicznej* (The Problem of Analogy between the theory of Evolution and the Theory of Socio-economic Formation). Warszawa/Poznań: PWN.

--- [1982]. The Theory of Development of Species and the Theory of Motion of Socioeconomic Formation. In *Social Classes, Action and Historical Materialism (Poznań Studies in the Philosophy of the Sciences and the Humanities*, 6). Amsterdam: Rodopi.

Lukács, G. [1923]. *History and Class Consciousness* (trans. from German). Cambridge: MIT Press 1968.

Mannheim, K. [1937]. *Ideology and Utopia. An Introduction to the Sociology of Knowledge*. New York: Harcourt 1977.

Nowak, L. [1975]. The Theory of Socio-Economic Formation as an Adaptive Theory. *Revolutionary World*, 14, Amsterdam: Gruener, 85-102.

--- [1977a]. *U podstaw dialektyki Marksowskiej* (Foundations of Marxian Dialectics. Towards a Categorial Interpretation). Warszawa: PWN.

--- [1977b]. On the Structure of Marxist Dialectics. In *Dialectical vs. Formal Logic*, *Erkenntnis*, **11**, 3, 341-363.

--- [1977c]. The Classic and the Essentialist Concepts of Truth. In (eds) M. deMay et al., *The Cognitive Viewpoint*. Ghent: Ghent Univ. Press.

--- [1979]. Historical Momentums and Historical Epochs. An attempt at a non-Marxian historical materialism. *Analyse und Kritik*, **1**, 60-76.

--- [1980]. *The Structure of Idealization. Towards a Systematic Interpretation of Marxian Idea of Science*. Dordrecht: Reidel.

--- [1983]. *Property and Power. Towards a non-Marxian Historical Materialism*. Dordrecht: Reidel.

--- [1984]. *Ani rewolucja ani ewolucja* (Neither Revolution nor Evolution). Frankfurt/M: Dialogue.

--- [1987]. A Model of Socialist Society. *Studies in Soviet Thought*, **34**, 1-55.

Quinton, A. [1973]. *Introduction*. In A. Quinton, ed., *Political Philosophy*. Oxford: Oxford Univ. Press 1973.

Weber, M. [1949]. *The Methodology of the Social Sciences*. Glencoe, Ill.: Free Press.

Poznań Studies in the Philosophy
of the Sciences and the Humanities
1991, Vol. 22, pp. 87–109

Piotr Buczkowski

REMARKS ON THE STRUCTURE OF SOCIAL CONSCIOUSNESS

1. Introduction

The purpose of this paper is an attempt to present an outline of a concept of social consciousness which would, on the one hand, meet the requirements that must be satisfied by a sociological theory and, on the other hand, make it possible to put forward hypotheses concerning the structure and organization of knowledge considered as an outfit of individuals and groups. It is the primary research area of the sociology of knowledge, encompassing the "dimensions" of individual consciousness which constitute social consciousness. The latter, it must be added, is not simply a set of all individual beliefs, for the category of social consciousness defined in this way is cognitively doubtful. At best it allows to register the state of consciousness of the members of a given group at a given time, but does not allow to make predictions concerning its changes or identify a typical group behavior caused by its state. Neither does the assumption that social consciousness is a non-individual construction existing in the form described by Durkheim and those inspired by his concept seem a satisfactory solution. The research techniques developed so far make it impossible to reconstruct such concept in a correct way. Ascribing a given group with a certain consciousness in an arbitrary way, no matter how spectacular heuristically within a given theoretical concept, has often led social scientists astray in contact with the live organism of social organization and the processes that characterize it.

I believe that the concept of social consciousness which functions theoretically must be embedded in real beliefs of members of particular groups. It must be embedded in such a way that these empirical references do not obscure explanatory or predictive potential on a non-individual scale. The present paper aims at offering an outline of a certain solution to that problem.

It seems trivial, at least for a sociologist, to claim that social functioning of certain groups of beliefs is closely connected with social roles. Society viewed as a structure in which a distinguished and at the same time primary characteristics is a specified system of roles can be analyzed from the point of view of recognition of both subjective and objective characteristics of these roles. That distinction refers to the characteristics resulting from an individual motivation of a particular role as one's own or a motivation resulting from the fact of group affiliation. Those two aspects do not always coexist. In atomized society an individual always loses the sense of group affiliation, thus also losing opportunity for a broader, non-individual definition of roles performed by that individual. The lack of sense of bonds cause the individual to be unable to verbally express values other than those which determine his individual activity. And *vice versa*: in an integrated society group affiliation determines individual behavior and the social dimension of activity acquires primary importance, thus crowding immediate individual interests out.

Social consciousness as a distinguished subject of research constitutes the primary domain of sociology of knowledge. One of the major subjects of interest within that science is the socially functioning ideology (or ideologies), which encompasses both the set of primary beliefs concerning social life and regulates at least some of the individual types of behavior. Therefore, I shall begin my considerations with a short presentation of the concept of ideology.

2. The structure of ideology

From the point of view of sociology of knowledge, in my understanding of its subject matter [Buczkowski 1986], the elements which determine performing certain social roles are: the *knowledge* possessed by the subject, the subject's *system of values* and the *system of norms* (rules of behavior), which guarantee on the basis of the possessed knowledge the realization of states of affairs preferred by individual while performing a given role. A subjective dimension of behavior is represented by those elements of knowledge, values and the system of norms which motivate individual activity and are perceived as such by individuals. On the other hand, objective dimensions of behavior are the ones which result from the fact of group affiliation. In other words, this distinction concerns the roles which are determined by the *interest of an individual,* which may — as such — remain in opposition (at least potentially) to the interest of other individuals, as well as the roles determined by a *group interest* which quite naturally assumes a certain unity of interests of a specified type. It is quite obvious that certain scopes of those two types of roles may and do overlap.

Empirical sociological research refers directly to individual consciousness. Only theoretical and reconstructive work makes it possible to conclude about the form of social consciousness, i.e. the primary subject matter of sociology of knowledge. Therefore, the question is how should one — acquainted with the empirical data referring to individual consciousness — reconstruct the consciousness of society or smaller social group. Standard questionnaires are intended by scientists to answer the question concerning the beliefs of subjects on certain matters, or — in other words — what is the specific fragment of their knowledge or systems of values. However, in my opinion the primary task of sociological reflection should lie in the question concerning the types of actions which individuals or social groups are apt to undertake in conditions perceived in a specific way. Question whether people have a definite attitude towards the reality around them is not the most important, at first we have to answer whether reality is apt to provoke certain types of behavior and what those types precisely are.

Yet, our primary theoretical task is to elaborate such a concept of social consciousness in which a distinguished role is performed by knowledge. It is knowledge that determines the scope of preferences, as preferences are defined on the real or possible states of affairs ascertained within it [Nowak 1976; Buczkowski, Nowak 1980]. It is also knowledge — both its conceptual and structural stratum — that determines which types of behavior are recognized as permissible and desirable. Therefore, knowledge is a distinguished element of consciousness and the sociologist is interested in the form of relation between individual systems of knowledge and what is referred to as social consciousness.

Both sociology and psychology (excluding the trends which derive from psychoanalysis) accept a belief that knowledge is one of the primary regulators of behavior. Moreover, majority of theoretical approaches assume as a starting point the theory of rational action, believing that individuals undertake such actions which — on the basis of their knowledge — lead to the realization of the state of affairs they maximally prefer [Kmita 1975]. A vital question is whether we, the scientists, are able to reconstruct that knowledge. Besides that, sociology also assumes that knowledge is — in most general terms — a set of beliefs of which individuals are aware. And yet, many actions — both individual and group — may be based upon knowledge which is not verbalized. Particularly common knowledge contains, above all, the rules of practical action without the consciousness of their "theoretical" background. Therefore, attempts at explaining certain types of behavior by investigating the verbalized beliefs may turn out to be declared because people are able merely to specify the aims of their actions or — in broader terms — values and, possibly, their relation with specific types of behavior. They are not always capable of indicating the relation to the possessed knowledge of the reality and, in specific situations, activate that knowledge. This cognitive "limitation" of individuals is analysed, for example, by sociolinguistics. I shall refer to its trend connected with Bernstein's [1972] theory of linguistic codes later.

According to that concept the verbalization of individual knowledge may be of a figurative ("restrictive" in Bernstein's terminology) or "intelectual" ("elaborative") nature. In most general terms, individuals using a certain type of linguistic code are able to indicate the arguments and sense of their beliefs concerning the reality or limit themselves merely to its figurative presentation, compatible with the way in which it is perceived directly (non-verbally). The problem is significant inasmuch as these beliefs refer to social reality and determine a good portion of behavior. It seems to concern particularly that class of beliefs which indicates a significant degree of universality within group consciousness, including the one conditioned ideologically. Researchers investigating that subject matter stress that the occurence of a specific type of language (linguistic code) allows for an analysis of both ideology and its social reception [Staniszkis 1985]. It is connected with a long perceived relation between beliefs typical of common consciousness and the contents spread by the means of mass communication. It does not mean, however, that primitive socialization, in the course of which an individual learns about the fundamental forms of categorization of reality, exerts no influence upon the reception of contents that are spread. It merely indicates that the mass media may, at least collaterally, organize group cognitive experience in such a way that they cause the occurrence of the desired types of behavior or attitudes on the part of those who control those media. Primitive socialization and education may thus be dominated by a style of thinking characteristic for ideology [Marody 1987]. Therefore, when reconstructing common knowledge, one must take into consideration the influence exerted by the existing ideology, and at least those of its elements which are determined by the current needs of the existing social order.

The individual's functioning in a group or — in broader terms — in a society requires the acceptance of such ideological contents which guarantee the preservation of the existing social order by performing the roles determined by the needs of a particular type of social order. It requires (and this is what indicates the ideological nature of that group of beliefs) that individuals perceive the reality and their place in it in the way desired by those who organize that order and that — in compliance with universal preferences — they evaluate specific states of affairs. That set of beliefs includes descriptive theses ascertaining the occurence of particular states of affairs and the value judgments referring to these states.

When reconstructing ideology, we should refer to social mechanisms that generate and select theoretical concepts [Buczkowski 1986]. I shall not go in detail so let me only indicate that the primary selectors are the systems of values accepted by the authors of concepts, paradigmatic requirements existing at a given time, and — what is most significant from the point of view of the needs of these considerations — ideological requirements put forward by the organizers of social order. Concept that aspires to be recognized as basic for ideological actions of propaganda must at the same time fulfill certain criteria irrespective of the current needs of the ruling group. It must have an internal structure which distinguishes between ideological concepts and *strictly* theoretical propositions. These criteria characterize all ideologies irrespective of their essential contents. Moreover, ideology directly uses a specific linguistic code (called *quasi*-elaborated by M. Marody [1987]), which is instrumentally subordinated to social actions.

Ideology may contain — similar to scientific theories — a certain hierarchization of factors and relations occuring between them. That hierarchization constitutes the basis for constructing descriptive statements and the ones which quasi-explain the social reality. However, contrary to scientific theories which aim at discovering the factors truly significant for a given group of phenomena, ideology reverses that order, indicating secondary factors [Nowak 1980]. The theses formulated on the basis of ideology describe reality in a falsified way, thus obscuring the true causes and course of phenomena. The internal structure of ideology respects at the same time one more property characteristic of scientific theories. The latter contain two types of statements: laws being descriptions of regularities (most significant relations) and concretizations indicating the modifications of these laws by taking into account secondary factors. Development of normal science consists in researchers' actions confronting the concretizations with experience and revising the repertoire of secondary factors, considering some of them as insignificant, while at the same time introducing new secondary factors, etc. They do not revise (at least until a paradigmatic breakthrough) the repertoire of primary factors, supporting them with *ad hoc* hypotheses. In ideology the functions of statements that describe regularities are performed by *stereotypes*, which — because of referring to insignificant factors considered to be the primary ones — are least susceptible to empirical verification.

There is also a clear relation between the structure of ideology viewed as a set of stereotypes and the remaining theses, on the one hand, and the system of values accepted by the organizers of social order, on the other. According to the concept assumed here [Nowak 1976; Buczkowski 1976], the system of values is also differentiated into primary and secondary values. Primary values occur as a rule in a concealed form, since they contain states of affairs connected directly with the realization of the interest of the ruling class. The system of values is spread in an incomplete form, encompassing above all secondary values, which are presented as the primary ones. Secondary values perform at the same time a role that is instrumentally subordinated in relation to the primary ones. Reversing the system of preferences — together with reversing the significance structure of factors and dependencies — determines the ideological dimension of the propagated contents. The last element of ideology is the norms that define the classes of actions which — on the basis of ideological knowledge — guarantee the realization of certain values.

A cognitively useful category is the concept of worldview. In most general terms worldview can be defined as a system which includes knowledge K about the reality (the knowledge about what exists and about the relations between the assumed beings), evaluations E characterizing the preferences which order the states of affairs, and norms N, i.e. classes of actions which guarantee the realization of desired values on the basis of a given knowledge.

Worldview understood in this way is an ideology only when: (1) K is a significance falsiy (in sense of Nowakowa [1976]), (2) E is an axiological reversal of the system of evaluations held spontaneously in a given society (primary values are presented as secondary and *vice versa*), (3) norms N assume such a significantly false knowledge K and axiologically reversed system of preferences N.

The presence of ideological norms, however, is manifested not only in the recommended or realized social actions. Besides these, there are also *quasi*-explanatory theses which ascertain the occurence of certain states of affairs. Those states and the relations between them are also evaluated from the point of view of ideology. The primary role here is performed by the language used in describing the reality. M. Marody writes:

> words denoting conceptual objects [...] possess at least two peculiar properties: (1) their conceptual contents or the relations between them and the designates are more 'arbitrary' (in the sense that they permit many interpretations) than in the case of words denoting physical objects; (2) they are not emotionally neutral, i.e. they are permanently connected with specific evaluations. The former of these possibilities enables, while the latter motivates manipulative actions in language, which aim at creating a certain image of the social reality (Marody [1982], pp. 109-110).

A consistent omission of certain categories, phenomena, etc. may indicate a concealed negative evaluation or opposition to the values included in ideology which are directly connected with the interest of the ruling class. That omission or failure to observe phenomena of a certain type is also connected with the above-mentioned twofold nature of the systems of values included in ideology. Because of the fact that primary values, i.e. the ones which do not include states of affairs most desired from the point of view of the interests of the ruling class, are not directly verbalized in ideology, the states of affairs connected with them (irrespective of their positive or negative "sign") can neither be directly revealed.

The image of reality included in ideology directly determines the rules of behavior. System of values mediates here between descriptive theses and norms:

> The scope of actions is, thus, determined by codes both in the dimension of purposes (differences of levels described in the knowledge about reality) and in respect of the course of those actions (various solutions according to the

level of description of reality). However, this scope is also pre-determined by the fact that linguistic codes, which define general cognitive abilities of an individual, make it more difficult to realize certain types of activity (*ibidem*, p. 113).

What I have said so far does not mean that individuals for whom a restricted or *quasi*-elaborated code is the primary tool of perceiving the reality are unable to verbalize the "theoretical" background of actions they undertake — which could be concluded from sociolinguistic considerations. The belief that certain social conditions favor the activation of various types of codes[1] seems to be more accurate. Referring to a psychological distinction between semantic and episodic memory [Tulving 1972][2], one can assume that consciousness is organized on two levels. The first one is operational level, i. e. knowledge used directly for the realization of particular types of behavior. It includes mostly rules of action which are consequences of the knowledge acquired earlier, existing in a latent form and updated, as it seems, whenever there are any incompatibilities between the predicted results of behavior and its actual consequences. The second level is that of latent knowledge, which is not updated currently, part of which is the individual philosophical perspective together with its inherent contents which justify the sense of undertaken actions, the sense which reaches beyond the single, everyday experience[3].

The action lacking cognitive reflection on sense of behavior is the peculiarity of ideology. It is ideology that imposes upon individuals and social groups certain types of behavior as a means of realization of universal values confining to indicate the relations between them, leaving aside the question of essential justification of these relations. It also seems that the problem of justification by referring to descriptive knowledge is of secondary importance from the point of view of ideology or, to be more precise, its social effectiveness, which exerts an influence above all on an emotional level. It results in the the fact that the same values or rules of behavior may be matched to various "theoretical" concepts, various descriptive systems.

In order to consider more closely the mutual relations between the analyzed elements of consciousness, it is necessary to indicate what the structure and organization of knowledge itself is. For that purpose I shall refer from now on to the psychological theory of cognitive patterns (Abelson, Shank [1977]; Trzebiński [1985]; Wojciszke [1986]).

3. The theory of patterns: an introductory attempt of a paraphrase

My interest in that theory will be limited. I shall examine whether it produces any direct implications for a sociological approach to knowledge understood as class of beliefs spread on a group scale. A paraphrase of the concept of patterns will be used for the purpose of developing the basic model of the concept presented in this paper. I shall leave the problem of its possible concretization for another occasion.

The basic question introduced by scientists dealing with the problems of consciousness is what mechanisms make it possible for an individual to select from among the multitude of environmental stimuli those elements which guarantee the effective action of that individual. The main hypothesis formulated on the basis of the theory of patterns is the statement concerning a peculiar organization of knowledge in the form of cognitive "systems" defined as patterns. According to that concept, "a pattern is a cognitive model, which — in relation to a specific area of reality — coerces our perception, thinking, memory, and also actions"(Trzebiński [1985], p. 264). In other words, patterns are cognitive procedures of a certain type, which — related to all ontological categories encountered by an individual (individuals, relations, events, etc.) — secure for that individual both the understanding of reality and its remembrance. Scientists dealing with these problems stress that patterns may be (and most often are) connected in a certain way with information about the emotional attitude of an individual towards a specific category and with indications concerning that individual's behavior in a given situation. In the light of previous considerations we can assume that schematically organized knowledge constitutes — together with the system of values and the system of norms — third system of individual consciousness.

Patterns include general information about typical representatives of a given ontological category, which are separated from the information with which an individual has been in contact. If I understand the theses of that concept correctly, patterns as representations of real categories would be their "model" expressions, allowing an individual to identify certain phenomena and — by being related to the system of values and the norms of action — to evaluate reality and react to it by means of applying appropriate types of behavior.

A problem that has not been solved is the mechanism of origination, as well as the structure of patterns. As far as the former question is concerned, two primary mechanisms are distinguished, which I would call the mechanisms of *creation* and *adaptation*. The creation (formation) of patterns consists in the use of procedures of generalization and selection of those aspects which are primary for a given category. A separate problem is whether one applies the strategy of idealization or takes into account the frequency of co-occurence of certain elements. I believe that the use of idealizational strategy, i.e. distinguishing aspects that are most significant for a given phenomenon, is the most common procedure. On the other hand, as it has been pointed out by Maruszewski [1985] in reference to hierarchical knowledge (Obuchowski [1970]), an individual applies a certain form of idealization (Maruszewski's "paraidealization") when forming cognitive structures which refer to his self-image and his view of the world. One can also have justified suspicions that patterns formed on the basis of idealization (or paraidealization) would be verbalized by individuals in a developed code. This supposition is supported by the fact that people are able to form representations of phenomena whenever the latter occur accidentally. We should add that it is not a complete, visual representation, but a formation of a general type (a semantic pattern of a higher order), which influences the classification and interpretation of similar phenomena, which can be included in a given pattern.

It is also supported by the second mechanism of adopting patterns, which I would call the adaptive mechanism and which is characteristic for the processes of socialization and education. In the course of socialization and education individuals often deal with ready patterns that are a result of group cultural experience. Socialization and education models also cause that people possess patterns formed on the basis of applying a cognitive strategy (characteristic for a developed code), without being able to reconstruct that strategy at the same time. To use the terms suggested by M. Marody, the functioning of patterns of that type can be explained with the help of a hypothesis of the quasi-developed code. Let us also notice that the socially occuring susceptibility to adopt patterns of that type (conditioned by socialization and education models) makes it possible to encode ideological beliefs which include simplified images of reality, not referring directly to the cognitive strategies on the basis of which they were formed.

Cognitive patterns are characterized by their typical internal structure, which respect two fundamental principles: the principle of prototypicality and hierarchy. "A system of specific values, most characteristic for a given specimen of a pattern, is called a prototype and constitutes a representation of an ideal specimen." (Wojciszke [1986], p. 43). The functioning of a prototype understood in this way is explained in various theoretical ways. The quoted author points out to three basic standpoints in that matter. According to the first approach, a prototype is a representation of a real specimen of a pattern; according to the second one, it is an average knowledge — taken from experience — about all encountered specimens; finally, the third approach is a thesis that a prototype is a system of properties of the highest distributive power (*ibidem*), selected from experience.

As can be noticed, all indicated approaches assume the observational nature of the properties of particular categories, what means that they must be accessible in an individual's direct experience. And yet, it is possible to point out to many patterns of a prototypical nature (e.g. from the area of religion, magic, rituals, or — above all — ideology) where at least certain properties are not of a directly observational nature. A classical prototypical pattern in our society is the pattern of a category of social property and a lot of sociological researches indicate an almost complete lack of empirical references in the individuals' judgements concerning that category. Therefore, I believe that the source of the problem of identification lies elsewhere. One of the primary criteria of distinguishing prototypical patterns from other ones could be — besides the degree of generality — the degree of their universality in group consciousness. In other words, prototypes would be the patterns which constitute part of group consciousness by the fact of their common acceptance within a given group. If this statement is true, then prototypes — after an appropriate transformation or without it — may be the regulators of group behavior. It does not mean that every updating of a prototype would be identical for every member of that group. Depending on values interfering at a given moment, various patterns of lower order could be "activated", thus generating various types of behavior, which — however — would remain within the class of types of behavior determined by that prototype. The degree of generality, universality of acceptance and the class of rules of behavior determined by that prototype would, therefore, be the criteria allowing to distinguish prototypes from among other cognitive patterns of individuals. With that approach we can maintain that — considering the varied degree of universality of values which differentiate social groups — prototypical patterns would be correlated with the primary values of social groups (in a specific case-classes).

98

The second principle of organization of patterns is their hierarchy. They are arranged in certain "trees", the roots of which are prototypes. If, however, it is true that prototypes may activate various subpatterns (patterns of lower order) of individual knowledge, which need not be common, then the question of primary importance is that of legitimacy of statements concerning the degree of universality of certain values or attitudes based upon the examination of prototypical beliefs (and such beliefs are most often observed in questionnaires). It may happen so that prototypes themselves activate different patterns of lower orders, which in turn may generate different rules of behavior. It is easy to notice that the same prototype P may activate (in a simplified example) both a pattern of a lower order S_1 and S_2, which in turn generate different rules of behavior, respectively R_1 or R_2, R_3 or R_4, which is presented at the table below. We shall neglect here what is going to be discussed later, namely that to each pattern, also the prototypical one, a class of behavior is ascribed.

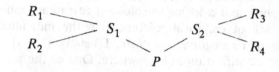

It may also happen so that individual or group knowledge includes only prototypical patterns concerning a particular fragment of social reality, being a result of propaganda influences. With the lack of rules of translation into a concrete (operational) knowledge, they remain "dead" and the pronouncement of their universality has an explanational value equal zero. If our considerations are accurate, we may assume that knowledge is arranged hierarchically into patterns of various orders. Between particular levels there occurs a relation of idealization (respectively concretization). Patterns of higher orders are constructs which neglect the characteristics less significant for a given class of phenomena. The roots of the structure of patterns are prototypes, which are the most idealized constructs. It must be stressed here that the concretization of prototypes into patterns of lower order, and those into patterns of a still lower order, may be based upon different properties (relations, etc.), characterizing specific phenomena. It causes the occurence on the same level of knowledge of individual different patterns concerning the same class of phenomena. These differences, however, do not refer to the primary characteristics included in prototypes. The analyzed type of relations between patterns provokes us to put forward a thesis that it seems more accurate to understand the relation between particular orders as a specific form of idealization, referred to as paraidealization (Maruszewski [1985]). It also seems that the type of pattern which is activated in the course of interpreting the reality depends upon the values accepted at a given moment. Therefore, the system of values is a factor facilitating both the activization of patterns and their concretization (respectively idealization) into a pattern of a different order. Let us now point to the fact that one of the mechanisms of change of patterns is the mechanism characterizing the change of ideological beliefs. In the situation of a clear incompatibility with experience, the changes of patterns take place on particular orders, i.e. the indicated incompatibility first causes transformations of patterns of lower levels and later patterns of higher levels. Meanwhile, the most permanent patterns are the prototypical ones. From that point of view one can suspect that particularly ideological stereotypes possess a prototypical form.

4. Structure and organization of consciousness: an introductory attempt at systematization

Let us now consider what types of relationships occur between the theoretical categories described above. The matter of primary significance for a scientist in this respect is the reconstruction of not only knowledge and values, but also indicating their relations with the rules of behavior for the purpose of making predictions about the direction of social activity of individuals. Reconstructing the relationship between cognitive categories of individuals, M. Marody proposes the following pattern:

language ⇒ perception ⇒ inference ⇒ images of reality ⇒ actions
⇓ ⇓ ⇓
actions   actions   actions

The first, 'vertical' type of interaction – says M. Marody – takes place when we use various ready elements of our cultural outfit – in the form of concepts, patterns of concluding, beliefs comprising visions of reality – in planning or realization of behavior. The other, 'horizontal' type of interaction reflects the active participation of language speaker in creating visions of reality, on the basis of which various types of behavior are realized (Marody [1982], p. 40).

In the light of previous remarks, the above pattern can be interpreted in the following way. On the level of relationships, they reflect the relation between knowledge and behavior, characterized by the activation of subsequent levels of organization of knowledge in the form of patterns of a constantly higher level. An individual perceiving specific phenomena interprets them on the basis of individual patterns that he possesses, which in turn activate patterns of higher levels, including the prototypical ones. That process of "intellectual" cultivation leads to forming a definition of a specific fragment of perceived reality, which in turn enables the individual to select a specific type of behavior. Let us point out that this process is directly connected with the subject's system of values. That connection may be of a twofold nature: on the one hand, an individual perceives the reality on the basis of specific values, while — on the other — these values perform a crucial role in selecting an appropriate type of behavior, which makes that realization possible. The second, "vertical" — in the author's terminology — type of interaction reflects the phenomenon of "nonverbal" behavior. This term means here that an individual does not carry out intellectual operations for the purpose of understanding the relations between images of reality and the selected types of behavior, but reacts in a somewhat behavioral way. In case of a relation of that type, the appearance of a specific pattern provokes a specific type of behavior. A vertical type of relation also describes the relationships considered above, i.e. the ones between images of reality, propagated in a prototypical form in ideology, and the required types of behavior, without initiating cognitive strategies. In other words, we can distinguish a class of types of behavior, whose characteristic feature is the lack of indirect links between prototypical beliefs and the undertaken actions.

However, in order to explain the nature of the analyzed relations in a more specific way, we must consider the remaining elements of consciousness (besides knowledge): the systems of values and norms (rules of action). From the point of view of sociology of knowledge, those elements are primary which constitute non-individual consciousness, i.e. *the ones which are shared by majority of individuals in a given group*, or — in most general terms — within the society as a whole. The basic theoretical assumption which is adopted here is the distinction that has been present ed in the sociological thought for a long time, namely the one concerning the distribution of social roles into organization and execution. On a general social order this distinction assumes a division into organizers of social order and citizens. A lot of empirical data — particularly the data collected in recent years — indicates that in our society this distinction is centered around subordination and domination of a political type. It is a matter of theoretical selection to form an introductory definition of the nature of that domination.

By and large, theoretical concepts can be divided into two groups. The first one includes concepts according to which the basis of social organization is comprised by the aspirations of the organizers of social order to obtain an agreement between various social groups constituting a global society by means of actions of a mediatory nature aimed at creating optimum possibilities of realizing the interests of those groups. The second group are the ones which assume a structural opposition of interests and indicate that the conflict resulting from it is the primary feature which characterizes the structure of a global society[4]. A structural conflict resulting from the monopoly of political decisions of one group can manifest itself on different levels of social organization (in the case of open disturbances, it encompasses majority of areas of social life). One of the most fundamental factors causing its occurence and promoting its existence is the degree to which particular groups are aware of its presence. In this respect a theoretical analysis reconstructing the state of social consciousness is cognitively useful not only for indicating its social conditionings, but also for understanding the nature of the conflict itself, its social reception and possible directions of development. Turner [1985], reconstructing the fundamental theses of the theory of conflict in Marxian version, wrote: "The more the subordinated social groups are able to develop common systems of beliefs, the more probable is that they may become conscious of their real group interests."

From the above point of view the primary category of describing consciousness is the relation between individuals (and groups) and the social system. This concerns not only elements which are common for specific social groups and are realized as such (Ziółkowski [1984]). This attitude towards the system also comprises the systems of values, attitudes towards reality which include its description, diagnosis, explanation and predictions of its development together with evaluations, as well as acceptable rules of behavior. The latter category can be divided into political types of behavior (i.e. the ones related to the social system) and those which lack this political dimension. Particular types of behavior may possess one or more dimensions, depending on the relations between them and values, as well as the system of knowledge (cognitive patterns activating specific attitudes).

However, if prototypical patterns are the ones which are characterized by a significant degree of universality in group consciousness, it seems natural to accept that they are correlated with the primary values of those groups (in a specific case of a dychotomous division of society — with class values). Primary values are determined by group interests (respectively class interests). Therefore, the types of behavior generated by prototypical patterns and class values can be defined as being *strictly political*. They are directly connected with the position of a given class in the social structure (its preservation or challenging), the realization of its interests, defining social differences, etc. This includes consciousness constructs, which have been defined above as worldviews. A worldview is also understood as an ideal type, which is typical of a certain group or class. If it meets the conditions defined in the second part of the present paper, it constitutes the ideology of a given group. However, the image of a society idealized in such a way requires a concretization.

According to the reconstruction of Marx's historical materialism assumed here, such a reconstruction may be carried out on the basis of the category of work division. Work division leads to the distinction within the global society of subsystems referred to as "historical moments" by the classical adherents of Marxism (Buczkowski, Klawiter, Nowak [1982]). Historical moments are, formally speaking, similar (in respect of their structure and the relations between them) to the global structure of a society. In each of them we may distinguish a type of material means peculiar for a given domain of social life, an institutional structure, and a type of consciousness. The category of work division allows for distinguishing of fundamental systems within a global society, i.e. the sphere of politics, economy, and spiritual production. These spheres are connected with specific relations of domination-subordination, which in different historical periods structuralize the global system. These historical moments of the lowest level in turn are divided into moments of higher levels. For example, within the economic moment particular areas of production are distinguished. Social consciousness relativized to particular momentary structures is comprised by group systems of norms, which also include specific knowledge, systems of values and rules of behavior, characteristic for a given domain (Buczkowski [1979]). It needs not be stressed that values are directly correlated with a group interest of a specific type.

In view of the fact that systems of group values are distinguished by preferences connected with interests (most often in conflict) of various groups, the actions which active people are apt to undertake may also directly refer to class values. If they are directly compatible or contradictory to class values, they remain — as I have already said — *strictly political* actions. In other words, if their objective results remain in opposition (or are compatible) to class interests, they may be qualified as belonging to a specific group of actions. However, if they refer merely *subjectively* (both through attitudes, declarations, or senses perceived socially), without remaining in a directly objective opposition to class interests, they may be referred to as being *quasi-political* (or politically marked).

It is not difficult to notice that political actions, as well as those politically marked, occur universally in periods when the existing order is challenged, i.e. in revolutionary periods, but also — what is quite natural — directly before and after such periods. As has been observed in 1980-81, in revolutionary periods all actions are politically marked, including the actions of *everyday adaptation* — the last class of types of behavior distinguished here. The everyday adaptive types of behavior, however, have no direct relation to specific class or group interests. They are usually determined by single individual patterns, which are connected with the rules of action, most often non-verbal, that are most easily updated. The latter means that an individual does not perform the operation of "interpretation" of a given pattern by applying a cognitive procedure and does not refer it to non-individual systems of values. To put it still differently, they do not force the application of cognitive procedures for the purpose of defining their sensibleness, referred to systems of values other than the individual ones. In the latter class of types of behavior we can distinguish two kinds: the first one is a group consisting of types of behavior resulting from real internalization of prototypical knowledge, i.e. the types of behavior proposed by the propaganda and based on an individual's belief that they lead to the realization of specific values. The second one includes types of behavior undertaken by individuals against their knowledge or values, and which are forced by the needs of adapting to the system or a threat of repression (although both types may be mutually related). These types of behavior are included in the class of such rules which an individual is able to interpret cognitively, although he does not do it because of the necessity of adapting to the requirements of the system or the reduction of a potential cognitive dissonance, as well as in the class of such rules which do not yield to interpretation because of the lack of appropriate cognitive patterns.

As I have stressed above, behavior is not directly connected with systems of knowledge, but it is mediated by the system of values. It is quite obvious that — besides cases of ideological exclusion of cognitive strategies — in order to realize the primary values it is necessary to make them instrumental. In other words, it is necessary to know the rules of translation of these values into instrumental values. This process of translation requires activating cognitive structures of a lower level. Then, instrumental values are translated into individual rules of their realization, i.e. they control the selection of specific individual types of behavior. The realization of values is connected with activating individual cognitive patterns. The level on which the primary values and their instrumentalizations occur together with the cognitive structures and rules of group behavior that accompany them can be defined as the *level of social consciousness*. It occurs along the *level of individual consciousness*, which encompasses individual cognitive structures, realizations of values, and the rules of action generated by them. Let us present it graphically, together with distinguishing the types of behavior discussed above.

the level of social consciousness	prototypes ⇓	⇒ primary values ⇓	⇒ political behavior ⇓
	patterns of lower levels ⇓	instrumental ⇒ values ⇓	quasi-political ⇒ behavior ⇓
the level of individual consciousness	individual specific patterns	realized ⇒ values	everyday adaptive ⇒ behavior

In the suggested approach vertical relations which occur between particular theoretical categories reflect the process of "intellectualization", or cognitive interpretation of reality. This process may occur in both directions at the same time. It corresponds to the theses occuring in the theory of patterns and concerning the possibility of activating the patterns "from above" or "from below". On the other hand, horizontal relations encompass the phenomena of "non-verbal" action, i.e. the ones lacking intellectual interpretation of true meanings of undertaken actions, or — in other words — without referring to cognitive strategies, which require each time to confront the rules of action with the knowledge possessed by individuals. Finally, diagonal relations encompass the class of types of behavior which are connected with the peculiar nature of influence exerted by ideology.

Let us consider the formal similarities between the elements of consciousness distinguished above. It seems that — similar to knowledge, organized in form of "trees" that spread from prototypes, through patterns of lower levels, to single individual patterns — the remaining elements are also organized in form of "trees". I have already stressed that in order to realize primary values individuals or social groups must translate these values into instrumental values, which are then operationalized in the way enabling them to undertake specific types of behavior. Such instrumentalization of primary values and operationalization of instrumental values must be based upon relations of concretization. Otherwise, the actions undertaken by individuals or groups will produce results other than the preferred ones. It seems that all cognitive limitations of individuals of which we have spoken may influence inadequate assimilation of values of lower levels.

However, in the model approach instrumental values are concretizations of primary values or idealizations (of a specific order) of operational values, i.e. the ones realized directly in actions. Similar to instrumental values, the primary values may be concretized in many different ways, depending on the position of a particular group in the social structure, the existing cultural configuration, or the possessed knowledge about the reality. Therefore, each primary value (if there are more than one) may be concretized into a class of instrumental values and those, in turn, into sets of realized values. It is evident that the structure of the system of values is formally similar to the structure of knowledge and the relations between particular elements of each of them are of the same type.

A similar situation concerns the structure of the system of norms, which is the third element of the proposed concept of consciousness. We can distinguish three basic types of actions: simple actions, complex actions, and actions of the highest level. We shall define simple actions as such in which no other actions instrumentally subordinated to them can be distinguished. In other words, a simple action is a set of factors the only sense of which is the operational value realized by an individual. The set of senses of simple actions understood in this way constitutes a condition necessary for the realization of an instrumental value of a given type. Because of the fact that instrumental value is relativized socially (rather than individually, as in the case of operational values), complex actions will include the objective results of simple actions undertaken by individuals. In this sense complex actions include — besides the sense determined by a given type of instrumental value — also the results of actions of a lower level, particularly simple actions. Because the indicated relation is defined by the results, rather than by socially ascribed senses, complex action is not a sum of simple actions. It is, therefore, possible to understand the relation between simple action and a complex one as a certain kind of idealization abstracting from specific actions which characterize simple actions.

To put it in simpler terms let us recall an example. The sense of working in a factory is to gain the means of subsistence, while its objective result is to produce a desired product as a result of cooperation. In still other terms, I use this distinction to emphasize the idea that complex actions may be based upon various simple actions, providing that individual realizations of particular types of behavior produce results of a specific type. Finally, complex actions of the highest level can be defined as systems containing the results of complex actions of the n-th level together with the sense determined by a specific primary value. It need not be added that the action of the highest level is not an instrumental component of any other action.

5. Conclusion

If the above considerations are an accurate reconstruction of the structure and organization of consciousness, it is possible to put forward certain methodological and substantial propositions for the empirical research conducted by sociologists. The primary task is, one can believe, not the examination of prototypical beliefs, which are revealed in standard questionnaires, but the analysis of the respondent's ability to concretize them. If it is justified to believe that these examinations offer us an insight merely into those "levels" of consciousness which are characterized by the existence of prototypes, then the hypotheses put forward on the basis of their reconstruction require a justification by indicating both the understanding of particular categories and the ability of their translation into patterns of lower level. This ability of translation may be analyzed not only by referring to the respondent's knowledge (which may be incomplete), but by referring to the systems of values and the systems of norms which the individuals are able to verbalize. It also seems that standard questionnaire examinations can be treated as a kind of pilot research serving the purpose of making hypotheses that must be verified with the use of other techniques such as a deep unhampered interview, minimizing the influence exerted by the interviewer and enabling the respondent to undertake a reflection over his own knowledge and the remaining elements of his consciousness. At the same time, I am aware of the fact that the proposed concept requires a detailed operationalization, although I conjuncture that the suggested directions of research may produce interesting cognitive results.

NOTES

[1] It is, therefore, possible to believe that a developed code is characterized by the fact of occuring in situations in which an individual undertakes an effort to identify his true group affiliation in the periods of disturbances of the order. Certain other factors influencing both the formulation of utterances and their understanding are indicated by Ziółkowski [1986]. One can also believe that in situations defined as "the paradox of stress" (Reykowski [1966]), individuals may activate those parts of their knowledge which remain unrealized in ordinary conditions and which enable them to use cognitive strategies and to verbalize beliefs in a developed code.

[2] I have only adopted the idea of this distinction, assuming that both "types" of memory have their equivalents in the organization of consciousness. It must be added that this distinction does not concern the stability, but rather – the direct application in everyday behavior. From this point of view updating of latent knowledge transforms it into operational knowledge, providing it regulates the behavior at a given time.

[3] This perspective may be understood as universal knowledge in the sense adopted by Obuchowski [1982].

[4] These are certainly not all possible types of order, although this division characterizes in my opinion a majority of concepts which are empirically operationalized and used for describing modern societies.

108

REFERENCES

Abelson R. P., R. Schank, [1977]. *Scripts, Plans, Goals and Understanding: An Inquiry into Human Knowledge Structures*. Hillsdale.

Bernstein B., [1972]. *A Sociolinguistic Approach to Social Learning*. In: (ed.) S. K. Ghosh, Man, *Language and Society. Contributions to the Sociology of Language*. The Hague/Paris.

Buczkowski P., [1976]. The Marxian Category of Burgeois Scientist, *Poznań Studies in the Philosophy of the Sciences and the Humanities*. 2, 1. Amsterdam: Grüner Publishing Co.

Buczkowski P., [1986]. Some remarks on sociology of knowledge. In: (eds.) P. Buczkowski, A. Klawiter, *Theories of Ideology and Ideology of Theories*. Amsterdam: Rodopi.

Buczkowski P., L. Nowak, [1980]. Werte und Gessellschaftsklassen. In: (Hrsg.) A. Honneth, U. Jaeggi, *Arbeit, Handlung, Normativität*. Frankfurt am Main: Suhrkamp Verlag.

Kmita J., [1975]. Humanistic Interpretation. *Poznań Studies in the Philosophy of the Sciences and the Humanities*. 1, no. 1. Amsterdam: Grüner Publishing Co.

Marody M., [1982]. Language and Common Knowledge (in Polish), *Studia Socjologiczne*, no. 3-4.

Marody M., [1987]. *Technologies of the Intellect*, (in Polish). Warsaw: PWN.

Maruszewski T., [1985]. Are the Idealizational Procedures used within the Scope of Common Sense Knowledge?. In: (ed.) J. Brzeziński, *Consciousness: Methodological and Psychological Approaches*. Amsterdam: Rodopi.

Nowak L., [1976]. Evaluation and Cognition, *Poznań Studies in the Philosophy of the Sciences and the Humanities*, 2, 1. Amsterdam: Grüner Publishing Co.

Nowak L., [1980]. *The Structure of Idealization. Towards a Systematic Interpretation of the Marxian Idea of Science*. Dordrecht/Boston/Lancaster: Reidel.

Nowakowa I., [1976]. Partial Truth – Truth – Absolute Truth, *Poznań Studies in the Philosophy of the Sciences and the Humanities*, 2, 4. Amsterdam: Grüner Publishing Co.

Obuchowski K., [1970]. *Orientation Codes and the Structure of Emotional Processes* (in Polish). Warsaw: PWN.

Reykowski J., [1966]. *Functioning of Personality in the Conditions of Psychological Stress* (in Polish). Warsaw: PWN.

Staniszkis J., [1985]. Types of Ideological Thinking (in Polish). In: *Krytyka*, no. 19-20. Warsaw.

Trzebiński J., [1985]. The Role of Cognitive Patterns in Social Behavior (in Polish), in: (ed.) M. Lewicka, *Psychology of Social Perception*. Warsaw: PWN.

Tulving E., [1972]. Episodic and Semantic Memory, in: (eds.) E.Tulving, W.Donaldson, *Organization and Memory*. New York.

Turner J., [1985]. *The Structure of a Sociological Theory* (in Polish). Warsaw: PWN.
Wojciszke B., [1986]. *The Theory of Social Patterns* (in Polish). Wrocław: Ossolineum.
Ziółkowski M., [1985]. Some Remarks on the Notion of Social Consciousness. In: (ed.)
      J. Brzeziński, *Consciousness: Methodological and Psychological Approaches.*
      Amsterdam: Rodopi.
Ziółkowski M., [1986]. Understanding Linguistic Utterances (in Polish), *Przegląd
      Socjologiczny*, vol. 33. Łódź: Ossolineum.

Poznań Studies in the Philosophy
of the Sciences and the Humanities
1991, Vol. 22, pp. 111-128

Andrzej Falkiewicz

THE INDIVIDUAL'S HORIZON AND VALUATION

The problems of sense and valuation are inseparably tied to that of freedom. No philosophical considerations about valuation and sense — the sense of human life, the sense of the world understood as extra-human existence — can provide satisfactory results unless the problem of man's freedom has been considered first. The latter has been one of the most difficult problems of philosophy. Due to this difficulty, most contemporary philosophers do not focus on the problem and think it should be considered within the context of people's "outlook on life"; this is one of the features of contemporary academic philosophizing. Owing to this universally used dodge, the word "freedom" is one of the most ambiguous words of human speech. We shall first make an attempt at explaining the essence of this ambiguity.

1

As every man, I desire to be free. But since I live among people I know that I should respect certain norms which provide for the same freedom for others. Hence, in whatever I do I am restricted by many prohibitions and commands. For example, as a car user I am compelled to observe a certain highway code which provides for the safety of others. At the same time I know, though it is only a hunch but an undeniable hunch, what servitude is. I experience servitude fully when I am imprisoned, when I have to comply with the severe rigour under the constant watch of the warden. This is a standard model and extreme situation. I apply to it specific situations experienced in life and I use this model to measure the extent of my servitude. And here I encounter the first difficulty. I notice that my dependence on the highway code and other norms of coexistence has little to do with servitude understood as above and that it cannot be measured with the model applied. It appears that the true problem lies elsewhere. Namely, even if I spend time in prison I can feel free, hence *be* inwardly free. On the other hand, as a member of society enjoying distinctions and endowed with numerous privileges I can feel enslaved and hence *be* enslaved. In this case I cannot be a fully objective judge of my own condition but I can easily recognize this state of bondage in others as I have experienced and I am constantly experiencing numerous pressures which can lead to the said bondage.

Thus, it appears that a distinction must be made between outward and inward freedom. The former indicates a certain extent of freedom restricted from outside and granted from outside. For example, social movements and doctrines postulating collective liberation always promise me freedom and in reality and technically this is a mere promise of a certain extent of outward liberty. Usually, this is liberty which is different and greater in its extent from the liberty I presently enjoy. It is a certain *great* extent of liberty... Great but not too great — politicians and other leaders of mankind, as well as educators and more cautious parents hasten to add. I do not want to elaborate on this extent; I only want to make clear that the issue *can* be discussed, that practically we are talking about an "extent" which, etymologically, implies a certain restriction, a certain diminution on the part of others, not myself.

In contradistinction to always incomplete outward freedom which I must understand as a certain area restricted by coercion, the notion of inward freedom is meaningful and important to me only if I understand it to its full dimension. I can agree that the achievement of complete inward freedom is very difficult or even impossible but I cannot justifiably accept that somebody else can restrict my freedom. Then I would be facing something opposite to freedom, i.e. servitude.

Here our language fails us. When we speak about freedom we usually think about liberty of action and its external limitations but it seems to us that this is enough. But liberty denotes lack of coercion. Freedom, on the other hand, is a certain inward attitude, spiritual disposition. External coercion is the opposite of liberty, whereas internal bondage is the opposite of freedom. It is true that coercion is often the cause of bondage. It is also true that external liberty and internal freedom are reciprocally related; the former can condition and facilitate the attainment of the latter. But this is not always the case. Often we experience the opposite, there is no simple relationship here — logically, psychologically and philosophically these are different matters.

These problems with terms, let us admit, deliberately committed by language, become easier if we consider the problem of human creativity. Every creativity, artistic, philosophical, scientific as well as creativity understood most generally and most broadly whose effects become apparent in the originality or authenticity of our everyday personal doings, is conditioned by inward freedom. We should accept that no creative act is fully free if the problem is considered from outside. The very superficial enumeration of the kinds of coercion under which a creative individual happens to occur, both consciously and unconsciously, could be the topic of a lengthy lecture. Every creator is bounded by the beliefs and convictions of his culture, the current knowledge about the world, the existing ideas about the goals and methods of his work, immanent standards peculiar at a given time to a given discipline and universally recognized as binding, the existing repertoire of forms of expression, a specific extent of evolution of genres — scientific, artistic, philosophical, literary... — well, in the case of literature, a writer is even enslaved by the current state of natural speech which decides what, whether and how he can express. Every creator also has some binding idea about what is expected of him and about what he must not do — this idea lies heavily on him when, and perhaps especially when, he does not want to fully meet the obligations imposed on him from outside.

Well, that's it. A creator is a creator to the extent to which he can go beyond what he has found in the world. A creative act is truly creative only to the extent to which it goes beyond an external coercion which is carried by the existing reality: when it goes beyond the achieved degree of collective consciousness, when it goes beyond dictates of common sense approved of by public opinion, when it goes beyond the current state of knowledge available to a given branch of science or art, when it goes beyond some generally shared norm, when it does not fulfill some expectation obvious to the contemporaries. We can express this more simply. A creative act is "creative" exactly to the extent to which it is free — and it can be defined only through its freedom. Someone overwhelmed by the obviousness of his neighbours' judgement and claims filed against him, overwhelmed by the wisdom and depth of the existing faith or knowledge (which, therefore, seem final), overwhelmed by the magnificence of the existing cultural achievements (which, therefore, seem uncrossable)... overwhelmed by all this and thus *enslaved* by all this, is not a creator. For these or other reasons such a man gives up his inward freedom, his actions replicate those of others and only imitate creativity in a less or more able way.

No conceptional subterfuge can undermine the definition adopted here. What is characteristic of a creator is a certain absolute novelty, courage of creating things and expressing judgements which have not been created or expressed — and only those are capable of doing this who have rendered themselves independent of the existing things and judgements. Who, in view of always powerful external coercion, knew how to get inward freedom.

And yet, contrary to what might result from the previous remarks, the extent to which human achievements are novel, is not infinite. And therefore the problem of freedom discussed here is still complicated. After all any novelty of a creator, if it is to be truly creative, must aim to be a "cultural value", for some collective human benefit. It is limited, *determined* by this benefit. In what way is it determined? How is it limited? What does then this independence, this inward freedom of a creator depend on? What additional conditions must be fulfilled by a creator so that he can achieve this free and at the same time creative frame of mind?

The answers are different. Here is some "shelled reptile". "A helm, a sailor, a ship — all in one". "The wave does not cling to him, nor does he cling to the wave". "No one knew his life, no one knows his ruin: They are egoists!" This picture and the unfavourable accompanying commentary have, as is known, been written by Mickiewicz in his *Ode to Youth* (Oda do młodości). Thus, we should infer that a true creator, or, more broadly, every free and consciously active man, should be an exact countrariety of this egoistic reptile... But to another great artist "it always seemed that it is something horrible to be a useful man". In his opinion, a creator is, in his nature, a dandy who "should live and sleep in front of a mirror". A "dandy does nothing. In any function there is something loutish. Can you imagine a dandy who harangues the crowd? Perhaps to mock". Dandyism is a kind of "cult of oneself which is more permanent than happiness that can be found in another person". "It is the pleasure of astounding and proud satisfaction that he who astounds is never astounded himself"... That was Baudelaire in his *Confidential Diaries*.

This divergence in the statements made by the two great artists astonishes. It cannot be explained by features peculiar to the two cultures; judgements similar to those made by Mickiewicz can be found in Victor Hugo, while Polish counterparts of Baudelaire's judgements can be found in representatives of the Młoda Polska (Young Poland) movement, if not earlier. This divergence cannot be explained by span of epochs since both answers were written at almost the same time; only one generation separates them. I will hasten to add that I do not believe any of the artists — in each one can find judgements which contradict those cited above. Perhaps the difference of one generation explains everything — when we talk about generations we usually place partial truths in opposition.

On the other hand, the whole truth is when a creator tries to arrive at independence and would like to be infinitely free (which is pointed to in *Confidential Diaries*), but also the creator should feel bound by something; something which is not him and which does not come from him (and this is what *Ode to Youth* says). I do not want to delay the true answer and I will reveal that what binds and what should bind a creator; what diminishes a creator-man but what at the same time augments his work, are values he cherishes (or better: he has internalized). Thus, we must ask what values are and what valuation is.

2

The shelled reptile, as defined by Mickiewicz, is an existence which is its own departure point, its tool and an aim in itself. In other words, it is its own horizon. *Echet Chufu*, "Cheops's horizon", is the name given by ancient Egyptians to his pyramid: the gigantic venture which took nearly a hundred thousand lives. But has it secured future life for Cheops himself? Highly doubtful. I am not concerned here with the validity of beliefs held by the Egyptians but with the fact that a body, buried with such an opulence and with pompous ceremony, was usually robbed and, having been damaged in this way, "ceased to live". Causes for the failure of such events are various. Egoist Cheops was not the only loser, but the very failure is obvious to us. Existence, we feel, is not and cannot be its own horizon. A sailor, a ship and the aim of the voyage are three different things. All our observations of ourselves and of what surrounds us confirm this. This is why a mother voluntarily giving her own life for that of her child is something we understand.

It is something we understand, yes... but must it necessarily be admirable? Naturalists would say that this heroic mother substantiates with her deed the preference of the species. Bearing this in mind we sometimes talk contemptuously about the "instinct" shared with the nature, about "maternal instinct", and with this explanation we try to neglect the sacrifice of that mother. This approach is not free from some philosophical absent-mindedness which deprives us of the contact with the existence as a whole, reduces human problems to the problem of anthropology and in this renders any serious philosophizing impossible. However, before I take up the main issue of my considerations, I would simply like to point out that there are also mothers without maternal instinct and at least for this reason the mother mentioned before deserves admiration. By the same token, there are better or worse fathers and the father's concern about the family is equally understandable and is explained in the same way. Well, though I am not sure, perhaps he is "more reasonable" and would not be as willing as the mother to sacrifice his life for that of the child.

Let us then consider the case of the reasonable father, an individual who sincerely and truly cares for his family but who would not be willing to sacrifice his life for it. This means that this man, without ceasing to be himself, accepts as his own the preferences of a *broader existence*. Yes, exactly and precisely so because the family with which he is emotionally bound, whose existence he is *part of*, is obviously his broader existence. In turn, a macrosocial group, some community of equally oppressed or privileged people with which that man identifies himself, is the broader existence of this family. And the people (nation) he is part of, if only he considers it his own, is the broader existence of this group. Europe, truly European culture or its values or, as we universally maintain today, Central Europe understood as a certain community of people who find themselves in the same geopolitical situation, can be the broader existence of that people (nation). And finally, mankind, that is the human species for which the heroic mother gave her life, is the broader existence of Europe, understood one way or another, irrespective of the fact whether and what intermediate existence we can find for ourselves (western civilization? Christian culture of the West?). The circle is finally complete.

But let us speak now about a reasonable man. Such a reasonable man, without ceasing to "reasonably" be himself, identifies himself, partly identifies himself, with broader existences. This means that to some extent he considers the interest of the broader existences of his own. In doing so, he unconsciously *values*, as valuation means rejection of one's personal horizon and placement of oneself in the horizons of one's broader existences. Such a man can be presented as a system of concentric circles whose centre is that "shelled reptile", the egoistic, egocentric self of the dandy, Cheops... of each of us. Now we understand what values are. Values are requirements of broader existences, emotionally our own, that is internalized, accepted as our own through less broad existences. Everyone who identifies their own interest with the interest of any of their broader existences (an individual — with the interest of the family or some broader interests, members of the family — with the interest of the macrogroup, people or some broader interests...) increases their spiritual area, that is attaches values. Valuation means placement of oneself in the horizons of broader existences and adoption of their preferences as one's own. This is what values are: requirements of broader existences which will be substantiated through less broad existences. Valuation occurs each time a less broad existence, "reasonably" substantiating its own preferences, at the same time substantiates requirements of the existences of which it is ... or better, of which it feels part.

Thus, I do not have to feel threatened with imprisonment to try to avoid knocking over a pedestrian each time I drive a car. It suffices that I feel "part of the mankind's existence", that I accept as mine the preference of the species, that I protect every human life. And therefore I do not feel that many of the norms I have to abide by — the observance of the highway code, of other norms of coexistence — which bind me and which I am supposed to observe, as an *external* limitation of my freedom. I limit my freedom myself, "inwardly", by means of those broader requirements I have internalized, by means of the values I cherish.

Internalized values are "my" world, the world I live in. The more spacious is the world, the more air there is in it and the better I feel. And quite contrary, the narrower the world, the stuffier it is. For example, when I focus only on myself and when I want, the shelled reptile, to become independent of the environment, of its prohibitions and norms that bind me, I perceive and feel only what restricts my Self. Paradoxically, precisely at this moment I am experiencing the violence of the world; striving for absolute freedom, I experience servitude. This is a mysterious feature of human, individual optics, its inalienable alchemy. Paying attention only to myself and wanting to make myself independent of *something*, I become slave to what I want to become independent of. Therefore any action attempting to establish "the only one and his property" is futile; various stunts of absolute freedom, of the sort of Max Stirner, de Sade, Leo Shestov (especially his early writings), put it into practice almost always with a suicidal result, physically or philosophically suicidal result. Only when I start going *towards something* and voluntarily become dependent of that; only when I inwardly identify myself with something greater than myself and when I accept that something's interests as mine, do I find the purpose in whatever I do and that purpose takes the attention away from my person and sets in motion my inventiveness. Only now do I find a "theme" which releases the inward freedom. Contrary to that other freedom, absolute but *negative*, freedom from something, freedom I strived for to no avail, only now do I experience *positive* freedom, that is creative freedom, freedom towards something. These problems are known, they have been extensively discussed in religious and philosophical literature, discussed by Luther, Hegel, Marx, Nietzsche. And although the problems have not exhausted the cause of human freedom, yet they can explain the paradox of the creator mentioned here. The paradox of novelty necessarily restricted, innovation necessarily conditioned, necessarily determined by something broader, which is the only true, culturally effective innovation.

Now the error of both answers, the one given by *Ode to Youth* and that given by *Confidential Diaries* becomes obvious. The artist to whom "the wave does not cling" and to which "he does not cling" condemns himself to useless freedom of the dandy, to that mere "pleasure of astounding and proud satisfaction that he who astounds is never astounded himself". But, in turn, somebody who would entirely rely on the wave gives up creativity and satisfies himself with echoing existing things and judgements, universal stereotypes, good tastes. On the one side we would have novelty for the novelty's sake, redundant and empty innovation of the expression; on the other — passive submission to public opinion, repetition of known truths, replication of somebody else's achievements. And in fact the hell of culture is paved with simulated creativity and uncontrollable innovation — and for the creator, for a truly free man, and hence truly creative, only a narrow path between them remains. The work of Witold Gombrowicz, generally appreciated but still not fully understood, consistent in its apparent inconsistencies, tells about such a narrow path. The path is the main theme of his work and, at the same time, the main object of the author's concern, discussed for the sake of the author.

Here are some well known words: "Monday — I. Tuesday — I. Wednesday — I. Thursday — I". This I (Self), full of conceit, repeated four times, begins Gombrowicz's Diaries. But in his *Diary 1953-1956* we also read the following: "I am like a tone in the orchestra which must come into tune with its sound, find a place of its own melody in the melody of the orchestra; or, like a dancer for whom not the dance is important but the desire to be together with others in dance. Hence, neither my thought, nor my feeling are truly free and my own; I think and feel 'for' people, to come in tune with them; I get biased as a result of this utmost necessity". Let us notice, the author speaks about the *supreme* necessity. Both his statements are written with unsurpassed precision and both are absolutely true. This, it would seem, unsolvable coincidence of independencies and dependencies, freedom and its contradiction in the name of the supreme necessity, recapitulates my reasoning and helps realize the components of the dilemma faced by the one who values.

3

An egoistic egocentric is free from this dilemma. This is obvious. But it is not known to the chauvinist, either. A chauvinist is somebody who fully identifies himself with only one of his broader existences. A family chauvinist substantiates the interest of his family by going, literally, over dead bodies. For a chauvinist of social liberation who solidarizes only with the oppressed (himself being oppressed as well), setting fire to the squire's mansion, killing a factory owner or hanging a communist will be the ultimate desire. I do not have to elaborate on the chauvinist of his own people as nationalism, that is national chauvinism, is probably best known and most broadly experienced. But we also have a chauvinist Europocentrism, today less frequently than before, and, why not!, chauvinist anthropocentrism, that "local patriotism" on the scale of mankind, responsible for drastic oversights in human thought and for irreparable harm done to the Earth's nature.

Thus, not every undertaking of the requirement of a broader existence is synonymous with valuation. The essence of valuation consists in the *mutual penetration of horizons*, in seeing the matter in such a way that horizons of broader existences, broader matters, are visible. The essence of valuation does not dissolve less broad horizons that "have been already crossed", including my individual horizon.

Because I, who am the author of all this, feel and think about it, must be somebody important also for myself, somebody worth thinking and feeling. After all, only under this condition can I feel a *valuating centre*, the spiritual focus of values. The very essence of valuation commands us to evaluate "reasonable" people, that is those who attach importance also to their own life, and non-fanatics, that is "normal" people, who admit a multitude of attitudes and, in their tolerance, are opposites of chauvinists.

This understanding of valuation naturally gives rise to a discussion about principles. The very question of extensiveness, that is objective position of specific existences, can be controversial. We are convinced about it by, for example, liberational dilemmas of the Polish socialism of the end of the 19th century, these obstinate controversies over the primacy of class and national issues in liberation. Today we can also discuss, and we do discuss, the primacy of Polish and Central European interests in our geopolitical situation. This is, in a way, a distant 20th century echo of that 19th century controversy over the hierarchy of issues, the order of action. And then it appears necessary to give priority to one of the existences and to decide what we consider truly important: social liberation or Great Poland or moral perfection or increased cognition or *tout court* of human prosperity, and what the latter would mean. Or perhaps we should consider something entirely else — and any such choice removes all others; any act of valuation entails, in its very essence, a controversy. However, because we normally value moderation, we value reasonable and normal people, then even such discussions are never too extreme.

Here, however, another difficulty arises. The following one. In everyday life we value moderation, we award bonuses to the moderate, but we generally venerate those "abnormal" and those "unreasonable" ones. We wish to lead a free discussion, we assume a multitude of attitudes and we try to avoid fanatics, but we usually put on the pedestal those who act otherwise. Do we not erect our monuments to those — leaders, discoverers, believers — who, just to spite the norm, have chosen only one cause and not very reasonably identified themselves with that cause so much that they were ready to sacrifice their lives for it? This convinces us that there is something to discuss. Most of what has been written in the course of centuries is just that discussion. I will not join this discussion here. In what follows I only want to explain why this discussion is led and why it will be led.

4

Let us consider that intimate impulse which induces people to valuate. It is the striving for sense. As a human being I am destined for sense. This means that I can consider *purposeful*, or sensible, if the word is understood narrowly, only that of my action which is compatible with *my personal reason for existing in the world*. This personal reason is that sense we look for, *the sense* understood broadly, philosophically, understood in the way which only philosophies dare to discuss. If I cannot find such a reason, if I am not able to realize it for myself, my life "will not have any sense", I will find myself in a "senseless world" and I will not be able to undertake any of my actions being fully convinced that what I do is right. It is this mode of deciding about my action which makes me different from other natural individuals known to me. This specific human desire, the striving for sense, founds culture.

The above answer is correct, I do not deny this. But such answers place me in a subjectively human horizon, within the circle of feelings which I must accept to be proper to us, human beings, and their specific articulations, which only we can understand. I would like to abandon this anthropocentric outlook of the issue, give a broader, extra-human meaning to the sense and human culture. And precisely in order to pass on to consider problems important from the viewpoint of existence I must first understand the problem of sense subjectively and functionally.

How does the striving for sense manifest itself in my private experience? Let us, for example, take my experience as the father. I am a loving parent and therefore I care for my children. All my emotional reasons, these "instinct" impulses, as we call them, speak for this care. But, all of a sudden, at some time, unexpectedly to myself, a question arises about the aim of my concern. A question arises "what for?" Well, I answer myself, so that my children could grow up and could bring up their own children and so that their children could bring up their own children, and so that their children could bring up their own children, and so on and so forth... If I continue repeating this, the care for my own children becomes more and more senseless. For a moment a thought comes to my rescue — I am going to bring up my children as righteous Poles. In my search for sense I come across this broader existence and, by abandoning my family, I endow my people with my feelings, with my "instinct" reasons. But by doing so I only delay the question about the aim of my actions. Well, is my concern supposed to serve some distant cause? So that in a thousand years (a thousand years is merely forty generations)... so that in a thousand years there is a white eagle with the crown on border stones (placed possibly far away from Warsaw) and the word 'Poland' appears on state documents? My feelings cool down — all this seems senseless to me. The situation will not be different if I emotionally identify myself with Central Europe and when I see in it the beginning of the Europe of Free Motherlands. The situation will not be different if I identify myself with my work as a humanist and I emotionally identify myself with the fate of my own species, that is when I bring up my children so that they are "noble people" and when I can think about mankind in the perspective of ten or twenty thousand years. Always at the very end of my thought my feelings will cool down — and this cooling down *makes me look further*.

It is as if I were to endow everything with feelings in order to understand their insufficiency. This is how I ascertain the phenomenon of sense "postponement". Sense is what always *appears to me in the horizon of the broader existence*, broader than that in which I am trying to find sense. I must go beyond this existence... I must go beyond *every* imaginable existence and reach some unimaginable existence which I cannot grasp, which is inaccessible to me. This existence is the end of my road. Please understand. Not only *can* I get there but I *must* get there; the logic of the sense looked for and always postponed inevitably takes me there. Man has been confirming this road from the early days of his history.

Every port I find will appear to me to be not ready, until I place it in something more spacious. My every action will seem pointless to me until I understand it in a broader context. Every perceived existence will seem incomplete to me, until I see it part of a broader existence. Destined to follow this course of feeling and thinking, each time I go beyond a certain specific situation and, backed up by emotional reasons, I get into a more general situation — into the horizon of that of the broader existences with which I was able to emotionally identify myself. This inevitably human road along which search for sense proceeds opens the gate to an *inconceivable area*, found beyond the broadest existence ever encountered, which I cannot control, which I cannot cognitively grasp. The name we give to this area, natural or supernatural, a proper name or no name at all, seems less important and is decided, to a large extent, on the basis of the inherited culture. For my present considerations is most important that the logic of the sense I am looking for and which is permanently postponed..., that this transition from "detailed" existences to more "general" ones inevitably takes me beyond my existential or human position and finally beyond my position on Earth and leads me into something "which cannot be understood", "which is unrecognizable". This is our ultimate, *natural* port which a human being existing in the world must finally notice. Writers of the Young Poland movement spoke about "nostalgia", Witkacy called this "insatiability", Master Eckhart spoke about "emptiness", Budda about "nirvana" and Martin Heidegger about the "experience of nothingness". Someone else might speak about the "temptation of the Absolute". But names are not important here. What is important is where I am going, what destination I must reach.

5

Every branch of culture satisfies its temptation or nostalgia in its own way; every type of creator goes beyond his earthly position differently. An artist, an eulogist of the physical beauty of the world, appeals to human senses. In the name of the highest vocation of his art he goes beyond the physics and reaches some eccentric, ecstatic reality. A metaphysicist goes beyond his earthly position differently — ... more precisely, he does not go beyond it at all, this is where he begins his road (the prefix meta- tells us about this) and tries to reconstruct the order of things, known to and experienced by us. To a contemporary scientist it seems that he does not go beyond the "existential" position since science, in his opinion, abandoned the support of philosophy and metaphysical guarantees long ago. Yet he does not notice that this presumptuous science, thrown on its own resources, contrary to its modern illusions, must always go beyond itself, must always reconsider its own foundations... and goals. A believer in God finds God, the Ultimate and Supreme Being, All-Embracing Love or the Only True Existence to which all other existences are emotionally, logically and philosophically subordinated... and therefore he would neglect roads penetrated by others.

Each branch of culture finds its *source of sense* in the broadest and the inconceivable. It finds this sense-generating factor which lets it convincingly and consciously undertake its own tasks. These foundations of religion, philosophy, science and art, these *most general situations*, differently conjectured, are always similar to each other. And yet, at a closer look, they are different. We can and should talk about this similarity as in it the essence of our culture is contained. But we must also notice the differences. Heidegger, talking about the metaphysicist and poet, said after Hoelderlin that they live next to each other on two mountain peaks but there is a gulf between them. A similar gulf separates the metaphysicist and the believer in God, a similar gulf separates the believer in God and the artist, the artist and the scientist... Differences must be noticed as in them there is the principle according to which specific branches of culture are differentiated.

Only in light of the above am I able to explain in what my reasoning is different from how we used to speak about values. Now when I have the culture I have the opportunity of *assigning pseudonyms to values*. I will say more — we *have* culture so that we can assign pseudonyms to values, so that we can transfer values in a form we can assimilate. From now on the culture itself is a value, Christianity is a value and values preached by Christianity and culture, such as good, beauty, truth, justice, are values. Clearly, we have here two things. The first is the process of recognizing something as *valuable*, worthy of sacrifice and respect, always controversial, evoking violent contradictions (for example, the process of transition from "the chosen people" to "mankind"); the other is the way in which values reached are *codified* (and then objects which are the carriers of values are also often called values). The former speaks about an intimately human way of approaching values, that is approaching valuation; the latter about the forms in which, in human eyes, "in the eyes of culture", these values are manifested. The former refers the creator to his radically private experience, the latter refers the follower to codified articles of faith, refers the recipient to texts.

My reasoning should not be understood to mean that professional creators, specialists of culture, are in some way specially privileged in their ultimate wishes and intentions. Every man finds such a most general situation, this extra-physical (natural or supernatural), this inconceivable area. At least every complete man, the one who wants to truly valuate and who knows how to truly valuate, that is one who knows how to internalize values, contrary to those who intercept values from the environment and, thinking about the environment, about the judgement of their neighbours, passively declare them. Precisely this inconceivable background seems to be the condition of true valuation.

What does it look like from the point of view of the one who values? Man, for example, is desirous of God and in this desire removes all obstacles in his way, removes everything which is not God. St. Augustine speaks about it beautifully in Book VIII of his *Confessions*, Buddists, trying to reveal, as they call it, "the true nature of the mind" also speak about it. But this desire, everybody notices this, removes obstacles and makes that the "earthly", practical aims referred to known existences are also formulated more easily and reached with a less effort. And although the full, mystic dimension of that desire is given only sometimes and only to a few, the desire, usually subconscious but deeply peculiar to man, seems to be the condition of every conscious action. It is what *places me in myself*, an indispensable factor determining the quality of my being in the world.

The choice between self love, love for family, love for motherland, love for every human being is still a choice but I can choose now more freely and with greater certainty. That inconceivable area helps me in my choice. It (I will never find out how extensive it is) emotionally relativizes areas I have already known and moves them in my feelings into some background; removes resistance I have so far experienced, duties imposed on me, choices made by others which I remember. Requirements which reach me from there (but I will never find out whether they will reach me and along which road) relativize, "cancel" to some extent these obstinate claims of existences I have recognized, always present in the surrounding world, announcing themselves with posters, slogans which I hear all to clearly. Only now and precisely for this reason can I conquer their importunacy; that is why I *can* choose between them.

It seems that this inconceivable area provides for this indispensable penetrability from which the horizons of existences I have recognized can draw. That *penetrability* is the condition of true valuation and permits speaking about the one who valuates (in former times he would be called God-fearing!). After all, the chauvinism of the nationalist consisted in the impenetrability of the national horizon. The chauvinism of the family — in the impenetrability of the family's horizon, the selfish egoism of the shelled reptile — in the impenetrability of the individual's horizon, etc. This penetrability of horizons explains why when I drive a car I go beyond the rationally understood concern about preservation of my own species and, moved by incomprehensible emotional impulse, I also try to avoid knocking down a dog. And when one thinks more deeply about this, conclusions drawn from this trivial example are not trivial at all.

6

So. Only when I introduce into my area that most general situation, this inconceivable area, do I valuate truly. I valuate irrespective of the slogans I hear and the pressure of the environment. I primarily choose either the family or social macrogroup or my people or my vocation of a humanist that is the cause of the entire species, or... Certain existences seem more important, others less important; a hierarchy is created. This hierarchy of existences, which in fact is a hierarchy of values, referred to myself, the only of the kind, becomes my personal reason for being in the world, my more or less ready *sense of life*. This sense decides about my work, makes it the aim or means of my action, explains choices I make, decides about my aspirations and ambitions. Only by stemming from my sense can I understand myself and my decisions, can I qualify them as reasonable or stupid. Only under this condition do I become a complete man.

This is *my* world. This is how this world originates and this is its structure. Above the actual and real world, which has the hierarchies that I perceive as "objective" and which I feel through the pressure of the environment, I build my own hierarchy which competes with the others, a certain future world, a world *wished for*, towards which I go in my feelings and thoughts; always first in feelings and then in thoughts.

Obviously this process should be seen more broadly and should be understood not only through my private experience. We all live as people in an objective world, a world which we can see as "objective". But we live in it so that we, people, could act in this world *culturally*, so that we could collectively create worlds symmetrical to the objective world and, in our opinion, equally capable of existing. For this purpose elements of the current world are placed in different configurations, they are given the wished for hierarchy and in this way some future and desired reality is created. These worlds, different from the objective one, these wished for hierarchies of values, are called senses. Now, these are cultural senses, peculiar to a given culture, cultural formation, social macrogroup, denominational group, etc.

In other words, in these collectively wished for worlds we find what I, the creator, struggle with every time — that which exists externally in the world. I always begin with and must begin with a ready sense contained in a given culture, which reaches me through the truths of science, art, philosophy, religion, shared by the contemporary. And, having begun in this way, inspired by these truths, in a way "controlled" by them, I valuate; that is I introduce transformations and corrections into the ready sense of a given culture, first group transformations and corrections, compatible with the views of my macrogroup, cultural formation, etc., and then individual transformations and corrections. Having begun in this way, and having ascertained a certain lack of sense in the existing world, I find my own sense in my private experience, I create my own hierarchy of values. In order to understand what is the discussion of principles and why this discussion is led and will be led, we must imagine a culture as a place in which such hierarchies, group and individual, infinite in number, meet.

7

In conclusion I return to the question of monuments which adorn our cities. The question is — why do we value "reasonable" people but erect monuments to "unreasonable" people? I think that in my private experience this looks as follows. If I have a personal reason to exist in the world, my own ready sense, that "well inhabited" world, created individually and with inner conviction, with a clearly defined hierarchy of existences, importance of problems, I am a man worthy of envy. Then I am "somebody", somebody who primarily is important for oneself. Values which I have internalized give me the internal advantage over those who only declare values in their concern about the judgement of the environment, which is always within the reach of others' looks.

For as long as I live under stable conditions, felt as "normal", my own world gives me spiritual comfort, peace of mind, inner happiness. It gives me the necessary distance to everyday problems and slogans heard, lets me oppose choices offered, facilitates insight into the worlds of others, built in the same way. In other words, it will give me the power of tolerance which we value in conversations. But this mild landscape of the soul changes when the keystone of my world — namely the value which I have considered supreme — be it the family, people, denominational group, social macrogroup or something else, is threatened with annihilation. In this situation, wishing to save myself, that is my world *with which I have fully identified* (this is the maximal degree of value internalization!), I must stake all on a chance. In defence of my self, every risk becomes purposeful and every sacrifice is reasonable. If I am primarily a parent, I must defend my children. If

128

I am primarily a patriot, I must defend my motherland. If I am primarily an artist, a scientist, a philosopher, a publicist, a believer, I should rather agree to die than renounce what I think is right. In ultimate situations it is the "reasonable" ones who fully identify themselves with the value they cherish, it is them who are ready for true rationality. The degree of the latter is measured with their inward freedom, with the power of their determination — with the magnitude of Antigone and Father Maximilian Kolbe — the magnitude of what, in defence of their sensible world, they managed to voluntarily give up. And, quite contrary, those who are not willing to make such a sacrifice must be considered enslaved and hence irrational. We accept their arguments, referring to "common sense", with a forced smile. And this forced smile suffices as an argument and justifies the above reasoning.

... And because these absolutely reasonable, truly reasonable people, are infrequent in extreme situations, we venerate them by erecting monuments. We venerate the free so that they could be reasonable; or we venerate the reasonable for their freedom. We speak then, quite justifiably, about feelings and, again quite justifiably, we use feelings, but we venerate *higher rationality*. We venerate a *rule*, infrequently adopted in individual life. However, our personal imperfectness blurs our general outlook and does not let us perceive this rule as a rule.

However, this rule of existence, also called freedom, and this degree of reason which goes beyond the human individual dimension and which makes us understand the world as reasonable, will not be discussed here.

Popowicka 108/3
50—238 Wrocław, Poland

Poznań Studies in the Philosophy
of the Sciences and the Humanities
1991, Vol. 22, pp. 129–171

Friedhelm Lövenich

HEILIGSPRECHUNG DES IMAGINÄREN
Das Imaginäre in Cornelius Castoriadis' Gesellschaftstheorie

Castoriadis' Buch ist getragen von der Motivation, dem Strukturalismus und dessen Subjektlosigkeit einen gesellschaftstheoretischen Entwurf entgegenzustellen, der das Subjektive in Gesellschaft und Geschichte nicht schon vom Ansatz her ausschaltet. Er formuliert deswegen die methodische Kritik am Strukturalismus, dieser verwechsele die Konstruktion einer Theorie der Gesellschaft und Geschichte mit deren Struktur selbst: "wer den Sinn als bloßes 'Ergebnis' der Differenz zwischen Zeichen betrachtet, macht aus den notwendigen Bedingungen einer *Lektüre* der Geschichte die zureichenden Bedingungen ihrer *Existenz*"[1]. Ein treffender methodischer Einwurf, der sich mit Adornos Kritik am Positivismus vergleichen läßt; mit Adorno teilt Castoriadis auch die Ablehnung des Systemcharakters von Philosophie und besteht ebenso wie dieser auf der Miteinbeziehung des Unbestimmbaren, Nicht-Identifizierbaren, das sich der "rationalistischen Illusion"[2] entzieht; er sieht ebenso wie Adorno die Unreduzierbarkeit der Individualität als Ärgernis des systematischen Denkens, der Theorie[3] sowie die theoretische Schwierigkeit, das Andere, Nicht-Identische überhaupt zu denken in einer Sprache der "Identitätslogik (...) Was diese Logik auszeichnet, ist der Umstand, daß sie eine wesentliche und unzerstörbare Dimension nicht nur der Sprache, sondern des gesamten Lebens und des gesellschaftlichen Handelns begründet. Sie ist selbst noch in solchen Diskursen am Werk, die sie umschreiben, relativieren oder gar in Frage stellen möchte"[4]. Aber was Adorno zur Negativität der Bestimmung führt, findet sich bei Castoriadis als Positivität und Positivierendes, Setzendes real im geschichtlichen Verlauf vor; was bei Adorno als Nicht-Identisches zur kritischen Spitze gegen das identifizierende Denken wird, gerät bei Castoriadis zur absoluten Schöpfung.

Denn Castoriadis will im Rahmen der 'Diktatur' des Symbolischen und der Struktur das Subjekt retten und führt so als konstitutiv für die

Gesellschaft ein Imaginäres ein, das — als prinzipiell produktives, kreatives und somit subjektives, subjekthaftes — in seiner gesellschaftstheoretischen Funktion gegen die Struktur gesetzt wird: er erhält so eine subjektiv-objektive, formend-geformte Ordnung der Gesellschaft und des Individuums — auf Kosten einer *Ontologie des Imaginären*, die letztlich der Überzeugungskraft seiner Argumentation schadet, da er das Imaginäre zwangsläufig, es wird noch zu sehen sein warum, als Unbedingtes konzipieren muß. Zwar denkt Castoriadis das Subjekt und die Gesellschaft immer als produzierend und reproduzierend zugleich: das Subjekt ist ein "produzierender Produzent"[5], die Gesellschaft "die spannungsvolle Einheit von instituierender und instituierter Gesellschaft, geschhenert und geschenhendert Geschichte"[6]; aber beim Versuch der Ehrenrettung des Subjekts muß er notwendig in seiner Konzeption den Symbolismus der Institution zugunsten des Imaginären theoretisch zurückdrängen und dieses stark in den Vordergrund rücken, zum Mittelpunkt der Geschichte machen.

Was Castoriadis eigentlich will, ist — um es mit einem von ihm angeführten Heideggerzitat zu sagen — " 'das Ende der Philosophie' und ihre 'Auflösung in die technisierten Wissenschaften' " verhindern[7]; dabei muß er sich jedoch, wenn er den Stil traditioneller Philosophie fortführen will, zwangsläufig ins Mythische verrennen.

Marx, Technik, Geschichte

Das beginnt schon bei seiner Marxkritik, mit der er sein Buch einleitet; die marxsche Theorie stellt für Castoriadis die Negativfolie dar, auf der er später, nachdem er ihre Unzulänglichkeit aufgewiesen hat, die gesellschaftliche Bedeutung des Imaginären eintragen kann. Castoriadis geht dabei von der richtigen Einsicht aus, daß es dem Kapitalismus gelungen ist, seine — laut Marx — immanenten Widersprüche, die zu seiner Aufhebung hätten treiben müssen, ebenso immanent zu lösen; anstatt nun aber zu einer marxistischen Neuinterpretation anzusetzen, verabschiedet Castoriadis Marx, indem er grundlegend seinen Ansatz angreift, "die Produktion, die durch Instrument und Gegenstände vermittelte Tätigkeit, das heißt die Arbeit auf 'Produktivkräfte' zu reduzieren, letzten Endes also auf die Technik, und dieser 'in letzter Instanz' eine autonome Entwicklung zuzusprechen"[8]. Es ist sehr fraglich, ob diese Reduktion von historischer Entwicklung auf Technikgeschichte der marxschen Theorie gerecht wird aber Castoriadis sieht genau diese Reduktion als Ursache für Marx' undialektische Mechanik gesellschaftlicher Verhältnisse an[9], der seine Theorie des Imaginären entgegenstehen soll, denn "die Geschichte kann nicht nach einem deterministischen Schema ... gedacht werden, weil sie der Bereich der *Schöpfung* ist"[10] — sie verläuft genialisch und unabhängig von Produktionsweisen. Das grundlegende methodische Problem Castoriadis' ist hier schon im Ansatz sichtbar: will er den Theorien der subjektlosen Rekonstruktion der Gesellschaft etwas entgegensetzen, ohne die grundlegende Intersubjektivität als gesellschaftliches Konstituens in Rechnung stellen zu wollen, dann muß er die imaginäre Allmacht des traditionellen Subjekts in wie auch immer verzerrter oder 'gesellschaftlicher' Form reinstituieren und zum Zentrum der Geschichte machen.

Gegen den unterstellten Determinismus und zur Bekräftigung des Schöpfungscharakters der Geschichte behauptet Castoriadis, daß der Kapitalismus nur fortexistieren kann, wenn er nicht etwa die Subjekte zu Rädern im Getriebe verdinglicht, sondern wenn es ihm gelingt, "ständig die im eigentlichen Sinne *menschliche* Tätigkeit der ihm unterworfenen Subjekte in Anspruch zu nehmen"[11], also ihre Subjektivität kapitalistisch zu organisieren; Castoriadis begreift das als die Aneignung des Imaginären als Schöpfung durch den Kapitalismus: die Entfremdung ist sozusagen, die enteignete, negierte Schöpfung. Das Imaginäre als Inkarnation des Subjektiven in der Geschichte wird damit trotz verbaler Einschränkungen — um die marxsche Theorie, die er irgendwie doch retten will, der Realität anzupassen — gekonnt aus dem Hut gezaubert als Erklärung für den Teil, für das subjektive Moment, das Marx' objektivistische Analyse übersieht und das hier als 'Einstellung', "Auffassung"[12], Weltanschauung auftritt. Es ist sicher nicht zu bestreiten, daß das eine immense Produktivkraft sein kann: nämlich das 'Sehen von Möglichkeiten oder eben auch 'Übersehen'; aber Castoriadis muß, um seiner Theorie treu zu bleiben, dieses Imaginäre als *creatio ex nihilo* konstruieren, die letztlich unabhängig ist von der gesellschaftlichen Ordnung, von der es nur eingeschränkt, nie aber produziert werden kann[13].

Selbst wenn die meisten Texte dieser Marxkritik bis zu den fünfziger und sechziger Jahren in die Zeit von *Socialisme ou Barbarie* zurückgehen und Abgrenzungen gegen den Stalinismus darstellen, so hat Castoriadis sie doch in dieser Form 1975 neu veröffentlicht; was Castoriadis da als 'Marxismus' aufbaut, ist nichts anderes ist als ein

Popanz, den er als idealistisch abqualifiziert, weil er monokausal denke[14] — ein Mangel, von dem er glaubt, ihn durch seine Theorie des Imaginären beseitigen zu können.

Ich werde mich daher im wesentlichen auf Castoriadis Konzeption dieses Imaginären konzentrieren und zu zeigen versuchen, daß diese zwangsläufig von ihrem Ansatz her ins Ontologische abrutschen muß, wenn Castoriadis das Imaginäre so auffaßt, daß das Gesellschaftliche lebensphilosophisch als dessen Erhärtung und Erstarrung erscheint, so daß die Gesellschaft als vergrößertes Individuum auftritt, das nach dem psychoanalytischen Modell der Psyche vorgestellt ist, und gesellschaftliche Intersubjektivität in Castoriadis' Gesellschaftstheorie durch diese *Psychisierung der Gesellschaft* keinen systematischen Ort finden kann, weshalb Gesellschaft eigentlich gar nicht gedacht werden kann.

Funktion, Institution, Entfremdung

Selbst wenn bei Castoriadis die Geschichte als nicht determinierbare stets durchs Imaginäre in Bewegung gehalten wird, bilden sich doch Stufen, Phasen und Epochen, die sich in Institutionen kristallisieren, aus; Institutionen sind kristallisiertes, geronnenes Imaginäres. Wenn diese Kristallisierungen festfrieren und unbeweglich werden, sind sie der Gesellschaft entfremdet; das Imaginäre ist daher genau dann der Gesellschaft entfremdet, wenn es aufhört, imaginär zu sein, und institutionell dysfunktional einfriert. Das scheint eine gute Ausgangsbasis für eine realistische Theorie der Entfremdung zu sein, denn gegen die — zum Beispiel lacansche — Ontologie, daß die Gesellschaft, Sprache durch ihren Charakter allgemeiner Ordnung *per se* Entfremdung des Individuellen sein muß kann Castoriadis festhalten, daß Entfremdung erst in der besonderen Form von dysfunktionalen Institutionen auftritt[15]. Um es in eine andere Theoriesprache zu übertragen: Castoriadis geht davon aus, daß Institutionen, Sitten und Normen die lebensweltliche Ordnung einer Gesellschaft zu ihrer Selbsterhaltung dienstbar machen, und scheint somit ebenso wie die habermassche Theorie der Moderne mit dem Konflikt von System und Lebenswelt zu arbeiten.

Aber eigentlich differenziert Castoriadis nicht zwischen System und Lebenswelt und ist nicht interessiert an einer Theorie der Moderne, die die Entfremdung in der modernen Gesellschaft angemessen interpretieren will, sondern er versucht, mithilfe des Institutionenbegriffs ein geschichtliches Gesetz für die Entfremdung überhaupt zu entwickeln, wobei die Entfremdung dann erscheint als das Gegenteil des Imaginären, des Subjektiven der Geschichte, nämlich als festgerostete Subjektivität. Da alles Imaginäre auch Schöpfung ist, ist Castoriadis' Kriterium für eine entfremdete Gesellschaft die bekannte Tatsache, daß sie sich selbst als Natur erscheint: "die Entfremdung oder Heteronomie der Gesellschaft ist eine Selbstentfremdung, bei der sich die Gesellschaft ihr eigenes Sein als Selbst-Institution und ihre wesenhafte Zeitlichkeit selbst verhüllt"[16]; was bei dieser heideggerisierenden Definition unklar bleibt, ist, daß diese Legitimation der Gesellschaft als Natur nur die ideologische Verschleierung realer Herrschaft ist, die nicht etwa erst die Folge der Verselbständigung von Institutionen ist, sondern die diese bewirkt. Der Grund für diese theoretische Verkennung von Castoriadis besteht in seiner Konzeption des Imaginären, die sich in der Formulierung auswirkt, die Gesellschaft entfremde sich von sich selbst, oder die Entfremdung einer Gesellschaft bestehe darin, daß sie "sich so instituiert, daß sie nicht sehen will, daß sie *sich instituiert*"[17]; daß sie ihren gesellschaftlich-imaginären Charakter leugnet. In seiner Theorie werden indirekt Gesellschaft und Imaginäres identifiziert, und die Gesellschaft erscheint zudem — darauf ist noch zurückzukommen — wie ein psychischer Apparat strukturiert, der seine Instanzen ausbildet und der von einem unbestimmten Grund, einem unbestimmbaren gesellschaftlichen Es angetrieben wird. Somit ist es auch kein Wunder, wenn Castoriadis Entfremdung analog zur Neurose vorstellt, in der die einzelnen Instanzen den psychischen Prozeß selbstzerstörerisch organisieren: die Entfremdung stellt sich für Castoriadis als die "Entfremdung der Gesellschaft von ihren Instanzen dar, als deren Verselbständigung *gegenüber der Gesellschaft*"[18]; die Gesellschaft und ihre Institutionen, die sie doch infiltrieren, die ihr ihre Struktur und Organisation geben, treten auseinander, ja gegeneinander. Das ist nicht zu verwechseln mit Habermas' System-Lebenswelt-Problematik, in der beide Bereiche als zur Gesellschaft gehörig begriffen werden, in der die moderne Gesellschaft als durch diesen Konflikt bestimmt angesehen wird; bei Castoriadis verschwindet nicht nur die Präzisierung dieser Konfliktgrenze, sondern der ganze Begriff von 'Gesellschaft' wird unklar und fragwürdig. Castoriadis ist nicht in der Lage, eine Definition von Gesellschaft zu erzielen, die diese Unklarheiten vermeidet, weil er generell für alle Gesellschaften sprechen will und deswegen eine Konzeption des Imaginären vorführt, die diese Problematik noch verschärft. Anstatt sich auf moderne Gesellschaften zu beschränken und diese zum Beispiel anhand der Differenzierung von System und Lebenswelt und den daraus erwachsenden Problemen zu analysieren — was ja bei seiner Bürokratiekritik durchaus nahegelegen hätte — konzipiert er einen Gegensatz von Gesellschaft als solcher und entfremdeten Institutionen, der sich einer demiurgischen Bestimmung des Imaginären als treibender gesellschaftlicher 'Kraft' verdankt; um sein Ziel zu realisieren, die Subjektivität in der Geschichte in Gestalt des Imaginären zu retten, muß er ein Konzept von Entfremdung vorstellen, in dem Herrschaft als gesellschaftliches Phänomen keinen systematischen Ort in der gesellschaftlichen Struktur finden kann.

Solange für die Entfremdung nur das Kriterium reiner Dysfunktionalität der Institutionen für die Gesellschaft gilt, solange besitzt Castoriadis keinen normativen Standard, an dem sich die Zerstörung und Entfremdung ablesen ließe, wie etwa an dem der Lebenswelt, obwohl er ja — in der Konstruktion des Imaginären — genau um diesen bemüht ist; das gleiche trifft zu, wenn er dies rein subjektiv, auf das einzelne Individuum bezogen zu entwickeln versucht, wobei er — und das ist, wie noch zu sehen sein wird, eine notwendige Folge seines Ansatzes — Gefahr läuft, einer 'individuellen Vergesellschaftung' das Wort zu reden. Laut Castoriadis wird die Entfremdung faßbar als die "Herrschaft eines verselbständigten Imaginären, das sich anmaßt, für das Subjekt die Realität und sein Begehren zu definieren"[19]; die entfremdete Institution erscheint als "Verselbständigung des Symbolismus"[20], der sich dem verselbständigten Imaginären verdankt. Castoriadis hätte es hier in der Hand gehabt, eine überzeugendere Analyse symbolischer Entfremdung: der Entfremdung im Symbolischen, zu begründen, denn für ihn ist diese institutionelle und symbolische Verselbständigung verursacht durch das Imaginäre, durch "die imaginäre Komponente jedes Symbols und aller Symbolik auf beliebiger Ebene"[21]. Es ist einsichtig, daß Castoriadis hier unterderhand eine andere Bedeutung für das Imaginäre, eine erheblich sinnvollere, wie ich meine, untergeschoben hat, so daß seine Definition von Entfremdung — daß Institutionen, das Symbolische dysfunktional werden — in diesem Licht sich der von Morris für die pathologischen Zeichen[22] oder der von Rossi-Landi für sprachliche Entfremdung angleicht[23], unter Umständen sogar verglichen werden kann mit Baudrillards Theorie der Simulation[24]. Aber Castoriadis muß diesen Vorteil verschenken, da diese Inanspruchnahme des Imaginären für eine Kritik seine Theorie des Imaginären zusammenfallen lassen würde, weil er nicht davon lassen will, sein Imaginäres als Schöpfung für die Gesellschaft konstitutiv sein zu lassen.

"Das instituierende Tun, die Selbstschöpfung der Gesellschaft"[25] bleibt also das Thema einer immer verzweifelter nach dem schöpferischen Geist der Gesellschaft suchenden Gesellschaftstheorie. Das Dilemma, in das sich Castoriadis' Begreifen des Imaginären begibt, zeigt sich im Nachdenken über die Institution als das Sichtbarwerden und Gestaltgewinnen des Imaginären; er widerspricht hier seiner vorherigen Definition der entfremdeten Institution als einer, die sich als Natur setzt und die die Tatsache ihrer Instituierung durch das Imaginäre vergessen machen will, wenn er über die Institution generell feststellt, "die Institution setze *sich* voraus; sie kann nur sein, als ob sie schon immer und ohne Einschränkung gewesen wäre — und als ob sie immer erst bevorstünde"[26]. Damit hat er den — im Sinne der soeben angesprochenen Bedeutung des Imaginären — imaginären Charakter formaler Organisationen angesprochen, aber den symbolischen Charakter von Institutionen allgemein gemeint; denn "diejenige Seinsweise..., in der die Institution auftritt: nämlich das Symbolische"[27], ist immer absolut, das heißt Institutionen sind immer legitimiert "als gesellschaftlich anerkannte Symbolsysteme. Ihre Leistung besteht darin, Symbole (Signifikanten) mit Signifikaten (Vorstellungen, Ordnungen, Geboten oder Anreizen, etwas zu tun oder zu lassen, Konsequenzen — also Bedeutungen im weitesten Sinne) zu verknüpfen und ihnen als solche Geltung zu verschaffen, das heißt diese Verknüpfung innerhalb der jeweiligen Gesellschaft oder Gruppe mehr oder weniger obligatorisch zu machen"[28]; sie sind als Institutionen ein reales und zugleich symbolisches Netz, das sich selbst sanktioniert"[29].

Dieser letzte Satz ist auf zwei mögliche Weisen verstehbar, die beide Castoriadis' Intention zuwiderlaufen: entweder spricht er hier vom Symbolischen des Strukturalismus, "des institutionellen Symbolismus und seiner relativen Autonomie gegenüber den Funktionen, die eine Institution erfüllt"[30], also von einer systemtheoretischen Legitimation durch Verfahren, die sich von einem lebensweltlichen Diskurszusammenhang über moralisch-praktische Fragen abgekoppelt hat — das aber widerspräche seinem Versuch, das Imaginäre als Subjekt der Geschichte zu installieren; oder er spricht vom Symbolischen der Institution in Form des festgefrorenen Imaginären als grundsätzliche

Entfremdung von der Gesellschaft, die — vom Symbolischen zwar
konstituiert — dieses aber nicht verändern kann, weil es die Entfaltung
des Imaginären blockiert — dies würde die reale Ohnmacht des
Imaginären gegenüber dem Symbolischen bedeuten.

Um diesem Dilemma zu entgehen, muß er das Imaginäre als
Schöpfung wieder ins Spiel bringen und besinnt sich auf dessen
instituierende, das heißt bildende, formende Fähigkeiten, denn "mit der
Institution der Gesellschaft wird jeweils ein Bedeutungsmagma
instituiert"[31], so daß dadurch das Symbolische als Ordnung des
Diskurses letztlich zurückgeführt ist auf ein Imaginäres als *creatio ex
nihilo*; Castoriadis kann somit sein Imaginäres die Funktion dessen
übernehmen lassen, was traditionellerweise das Symbolische genannt
wird.

Symbolisches als Imaginäres

Funktion und Leistung des Symbolischen ist die 'Herstellung' von
'Realität' in einer bestimmten Gesellschaft; das Symbolische ist die
'Ordnung des Diskurses'[32], die bestimmt, was in einer bestimmten
Gesellschaft als 'Welt' zu gelten hat, die den Objektbereich definiert,
auf den sich die Subjekte beziehen und den jeweiligen Modus dieser
Beziehung; erst durch diese Definition von Realität entsteht aus den
Individuen eine Gesellschaft, wobei das Verhältnis hier das der
Gleichursprünglichkeit von Gesellschaft und Deutung von Welt ist.

Bei Castoriadis erscheint dieses Symbolische als Imaginäres, denn er
stellt fest, daß Naturtatsachen, wie das Geschlecht, gesellschaftlich
umgeformt werden und daß sie "ins Netzwerk der gesellschaftlichen
imaginären Bedeutungen eingehen, umgearbeitet, umgeschmolzen und
im Wesen verändert"[33] werden, da sie "auf das Magma sämtlicher
imaginärer Bedeutungen der betreffenden Geselllschaft"[34] treffen. "Die
Institution der Gesellschaft ist Institution einer Welt von Bedeutungen,
die als solche offensichtlich eine Schöpfung, eine jeweils besondere
Schöpfung ist"[35] — eben das Symbolische; das Symbolische ist das, was
aus einer amorphen Masse diffuser Wahrnehmungen eine geordnete
Welt macht, indem es durch Identifizierung und Differenzierung
Bezeichnungen einführt, die mit Bedeutungen belegt sind; genaugenommen gibt es keine Welt, solange es kein symbolisches Muster gibt,
sie zu erkennen; 'wahrnehmen' heißt, diesem unbewußten Muster
folgen[36].

Aber obwohl Castoriadis — außer der Beteuerung, die Gesellschaft sei schöpferisch — genau dasselbe sagt wie der Strukturalismus, nämlich daß es gesellschaftlich instituierte Strukturen gibt, versucht er, dies herunterzuspielen, um seinen Begriff des Imaginären — der dasselbe meint wie das Symbolische — zu legitimieren; denn "ebenso wie die Idee der 'Struktur' in der Ethnologie und Soziologie lassen das 'Gesetz' und das 'Symbolische' die *instituierende* Gesellschaft verblassen", wie er meint[37]. Dieser Hinweis auf den Akt der Instituierung ist ihm wichtig, weil er ja der Abhängigkeit vom Symbolischen den Kampf angesagt hat durch das Imaginäre, denn "unser Verhältnis zum Gesellschaftlichen — und zum Geschichtlichen als dessen Entfaltung in der Zeit — läßt sich nicht als Abhängigkeit beschreiben, das ergäbe keinen Sinn. Es ist vielmehr eine *Inhärenz*, die als solche weder Freiheit noch Entfremdung bedeutet, sondern auf deren Boden Freiheit und Entfremdung überhaupt erst möglich werden"[38]; in den Bedingungen dieser Inhärenz sieht Castoriadis die "gesellschaftlich-geschichtliche Institution der Dinge und einer Welt (...als) Husserls Lebenswelt"[39] verkörpert. Wie Lacan faßt auch Castoriadis die sozialisatorische Schranke der Einführung ins Symbolische — das bei ihm eben Imaginäres heißt — als den Ödipuskomplex[40] auf, wo das Unbewußte erst entsteht und in dem sich die 'Inhärenz' festsetzt; er interpretiert ebenso wie Lacan das Unbewußte als Diskurs des Anderen, als gesellschaftlichen Ort im Individuum, der laut Castoriadis vom Imaginären charakterisiert ist[41]. Castoriadis sieht diesen gesellschaftlichen 'Inhalt' im Subjekt als Voraussetzung von dessen Tätigkeit überhaupt an; bleibt es diesem 'verfallen' — wie Adorno gesagt haben würde — so bleibt es heteronom und entfremdet; gelingt ihm eine bewußtseinsmäßige Distanz dazu, ist dies eine Autonomie[42], die strukturell aufbaut auf der sich ständig bewegenden Struktur des Imaginären, seinem Magma, das sich in der Beschreibung von Castoriadis als identisch mit Lacans 'Tresor der Signifikanten' und ihrem ständigen Gleiten erweist, denn "in diesem Magma gibt es verhältnismäßig breite Ströme, Knotenpunkte, eher klare und eher düstere Zonen, Gesteinsbrocken im flüssigen Ganzen. Aber das Magma kommt zu keinem Stillstand, hebt und senkt sich unablässig, verflüssigt das Feste und verfestigt, was beinahe nichts war. Und wenn sich der Mensch im und durch den Diskurs bewegen und schöpfen kann, wenn er nicht auf ewig an die eindeutigen und festen Signifikate der Wörter, die er verwendet, gefesselt ist; anders gesagt, wenn die Sprache ist, was sie ist, dann eben wegen der Beschaffenheit des Magmas"[43]. Es ist also — hier allerdings im Gegensatz zu Lacan — nicht die Sprache selbst, der Diskurs, der Ordnung und Veränderung bewirkt, sondern 'irgendetwas' hinter der Sprache, unter den Wörtern, jenseits der Ordnung, ein wabernder Kern der Sprache, der in seiner Unstrukturiertheit der jeweiligen symbolischen Ordnung immer entgegensteht: das imaginäre Magma, das bei Castoriadis noch das Symbolische umfaßt.

Castoriadis grenzt sich ab von Lacans Symbolischem als Begriff für das, was er das Imaginäre nennt, indem er einfach behauptet, "der Begriff des 'Symbolischen', wie er in Frankreich von gewissen psychoanalytischen Strömungen verwandt wird, entspricht in Wirklich-keit einer Komponente bestimmter gesellschaftlicher imaginärer Bedeutungen, nämlich dem Aspekt ihrer instituierten Normativität"[44]; damit hat er nichts anderes getan, als Symbolisches und Imaginäres einfach zu vertauschen und die Begriffe einfach auszuwechseln ohne Grund, und hat somit das Symbolische auf "die Beziehung zwischen der Bedeutung und ihrem Träger (den Bildern oder Figuren)"[45], also auf die rein signifikative Funktion des Zeichens reduziert. Das Symbolische als bloßes Zeichen verkommt so zur reinen Bezeichnung, zur symbolischen Ebene, auf deren Basis wir sprechen; seine bedeutungs- und damit weltkonstituierende Rolle — die symbolische Funktion, die mit der Gesellschaftsstruktur vermittelt ist — geht dabei verloren.

Das ist nicht bloß ein Streit um Worte, denn wenn Castoriadis das Symbolische austauscht gegen das Imaginäre, so gelingt es ihm dadurch natürich auch, das Symbolische so zu definieren, so zu konstruieren, wie er das Imaginäre begreift: als unmittelbare und permanente Schöpfung; dies aber ist eine Assoziation, die sich mit dem Begriff des Symbolischen schlecht verbindet. Er muß also die Begriffe austauschen; damit hat er zugleich die real vorhandene symbolische Dimension der Gesellschaft als imaginäre und geschöpfte definiert im Sinn seines Verständnisses von Schöpfung, eines gesellschaftlichen 'Willens': Setzung. Das Ergebnis ist, daß, wenn die Gesellschaft als Bedingung und Bedingtes dieser 'Setzung' erscheint, sie eine mythische Gestalt erhält, aus der als amorpher Masse 'irgendwie' eine im Urknall "mit einem Schlag"[46] strukturierte Ordnung plötzlich da ist.

Castoriadis geht tatsächlich davon aus, "daß das Gesellschaftliche (oder Geschichtliche) als wesentlichen Bestandteil Nicht-Kausales enthält (...) Das Nicht-Kausale... erscheint nicht nur als 'unvorherseh-bares', sondern als *schöpferisches* Verhalten (der Individuen, Gruppen,

Klassen, ganzer Gesellschaften); nicht bloß als Abweichung von einem bestehenden, sondern als *Setzung* eines neuen Verhaltenstyps; als Institution einer neuen gesellschaftlichen Regel, *Erfindung* eines neuen Gegenstands oder einer neuen Form"[47]. Dieses Nicht-Kausale ist nun natürlich "das eine Schöpfung *ex nihilo* ist (...) Das Imaginäre, von dem ich spreche, ist kein Bild *von*. Es ist unaufhörliche und (gesellschaftlich-geschichtlich und psychisch) wesentlich *indeterminierte* Schöpfung von Gestalten/Formen/Bildern, die jeder Rede *von* 'etwas' zugrunde liegen"[48]; es ist also die Basis jener Inhärenz, die den Subjekten als Hintergrund gegeben ist, der symbolischen Lebenswelt.

Castoriadis interessiert sich nun am "Rätsel der Kohärenz"[49] dieser Inhärenz, dieser symbolischen Ordnung von Gesellschaften, also an dem, was 'das Gesellschaftliche' eigentlich ist; da er einen Determinismus — die Monokausalität — ablehnt, gelangt er dazu, "die Verkettung der Bedeutungen"[50] zu attestieren, die den Rahmen der gesellschaftlichen "Totalität"[51] ausmachen und "eine organisierende Struktur, die einen Aspekt jener Totalität ausmacht, aber noch etwas anderes ist"[52] — ein Ansatz, der wieder zurückführen könnte auf die Struktur des Symbolischen. Aber Castoriadis konzipiert diese Kohärenz inhaltlich als einen "einheitlichen Sinn..., der sich behauptet und durchsetzt"[53] und den Zusammenhang der gesellschaftlichen Formen als 'Gesellschaft' bezeugt:

was dieser Vielheit (von Bedeutungen – FL) individueller Vorstellungen (und vorstellender Individuen) Zusammenhalt verleiht, wäre demnach als gesellschaftlich-geschichtliches phantasma zu bezeichnen, als 'gesellschaftliche Wortvorstellung' eines bestimmten Wortes in seiner abstrakt-materiellen, von seiner Beziehung zur Bedeutung ganz unabhängigen Existenz. Gesellschaftlich ist diese Vorstellung, da sie allen und keinem, niemand Bestimmtem gegenwärtig ist. Dieses gesellschaftliche *phantasma* geht nicht in den Schemata auf, in denen man die Imagination und das Imaginäre stets hat denken wollen und nicht hat denken können. Es ist offensichtlich weder abgeschwächte Wiederholung, Reproduktion oder partielle Retention eines Gegebenen, noch ist es Nachahmung oder Ähnliches. Es ist vielmehr Schöpfung, Setzung, Institution einer nicht-realen Figur oder Gruppe von Figuren durch das gesellschaftliche Imaginäre; eine Schöpfung, die konkrete Figuren (die Materialisierungen, die vorkommenden Exemplare des 'Wortbildes') als das sein *läßt*, was sie sind: Wortgestalten, Zeichen (und nicht Geräusch oder Spuren). Imaginär: unmotivierte Schöpfung, die sich nur in der und vermittels der Setzung von Bildern vollzieht. Gesellschaftlich: undenkbar als Werk oder Produkt eines Einzelnen oder einer Masse von Individuen (das Individuum selbst ist gesellschaftliche Institution), weder aus der Psyche ableitbar noch erklärbar[54].

Diese imaginäre Institution einer nicht-realen Figur ist jedoch real im Sinn von gesellschaftlich 'gültig' und wirksam, also in dem Sinn, daß sie eine Bedeutung hat[55], sogar die zentrale Bedeutung dieser Gesellschaft ist, die noch von "sekundärem Imaginärem" umgeben ist, so daß das "zentrale Imaginäre"[56] aus dem Bewußtsein verschwindet, zur den Subjekten unbewußten gesellschaftlichen Struktur wird. Das Imaginäre wird somit geschichtlich radikal, Wurzel jeder gesellschaftlichen Ordnung.

Das wesentliche Argument zur Rechtfertigung dieser gesellschaftlichen Radikalität des Imaginären seiner Unbedingtheit, Unabhängigkeit und Spontaneität, kurz: seiner Ursprünglichkeit ist, daß dieselben, zum Beispiel biologischen, Tatsachen in verschiedenen Gesellschaften so unterschiedliche Formen der gesellschaftlichen Instituierung produziert haben, was sich Castoriadis nur mit einem radikalen Imaginären erklären kann, denn "die Wahl der Punkte, derer sich der Symbolismus bemächtigt, um die Materie mit Heiligem zu beladen und somit noch einmal zu 'heiligen', scheint größtenteils (aber nicht immer) arbiträr zu sein"[57]. In dieser Arbitrarität, dieser Willkürlichkeit der Beziehungen von Zeichen und Bedeutung — die Saussure als konstitutiv für die Beziehung von Signifikant und Signifikat festhielt — sieht Castoriadis das Imaginäre verkörpert: es *ist* diese Willkür der Herstellung gesellschaftlicher Bedeutungen: darauf daß er damit nur einen anderen Begriff für die — immer nur, nämlich zur Gesellschaftsstruktur, relative — Arbitrarität des Symbolischen einsetzt, ist schon hingewiesen worden; daß Castoriadis hiermit letztlich genau den theoretischen Standort des Strukturalismus, den er ja 'angreift', einnehmen muß, ist noch kurz zu zeigen, sowie die Tatsache, daß er damit auch die Konsequenzen des strukturalistischen Ansatzes übernimmt.

Heiligsprechung des Imaginären

Im radikal Imaginären verortet Castoriadis die 'Geschichtlichkeit' der Gesellschaften: indem sie Institutionen und damit gefrorenes Imaginäres sind, in deren Rahmen Subjekte ausgebildet und hineingebildet werden, sind ihre Formen der Integration rein instrumentell verursacht, weil abhängig von der gesellschaftlich definierten Realität: den imaginären Bedeutungen, dem Imaginären der jeweiligen Gesellschaft, nach dem sie sich 'sein lassen', organisiert und instituiert. Das

Imaginäre erscheint somit als 'Basis' der Gesellschaft, Motor der Geschichte; das Imaginäre als Konstitution der Realität gewinnt so den Charakter einer unmittelbaren Produktivkraft. Aber genaugenommen kann Castoriadis mit seinem Konzept die geschichtliche Evolution nicht erklären, denn er archaisiert das Imaginäre, indem er es zur arché, zum Ursprung des Gesellschaftlichen macht; diese überzeitliche Heiligung nimmt dem Imaginären jedoch den historischen Charakter: das Imaginäre wird zum Sakrament in Castoriadis' Gesellschaftstheorie.

Castoriadis erwähnt an einer Stelle[58] den Begriff der 'positiven Rationalität', den er analog zum positiven Recht begreift und der die einer bestimmten Gesellschaft zur Verfügung stehende Rationalität bezeichnet, etwa vergleichbar mit 'Paradigma', und der unterschieden ist von Webers idealtypischer Rationalität; Krisen treten laut Castoriadis dann auf, wenn die positive Rationalität einer Gesellschaft nicht mehr zur Steuerung ausreicht, also ein 'Paradigmenwechsel' — auch in gesellschaftlich weniger bedeutenden Bereichen — notwendig wird. Unverständlich, genauer: unverstanden bleibt im Rahmen einer Archaik des Imaginären in seiner Theorie jedoch, wie es zu einer solchen Krise überhaupt kommen kann. Hatte Marx noch Krisen dieser Art — im Bereich der Legitimation, im 'Bewußtsein' einer Gesellschaft, das heißt im Symbolischen: auf der Ebene der Weltdeutung und Selbstinterpretation einer Gesellschaft, die ihre Selbststeuerung ermöglicht — gesehen als Folge von Veränderungen an der 'Basis' der Gesellschaft, die vom herrschenden Legitimationsparadigma nicht mehr aufgefangen werden können, so bewirkt Castoriadis' Fassung des Symbolischen als 'Imaginäres', als nicht abgeleitetes, sondern archaisches, daß unklar wird, warum es jemals infragegestellt werden kann; warum es überhaupt 'heiße Kulturen' und nicht etwa nur 'kalte' wie die des Mythos gibt.

Die Antwort von Castoriadis liegt in seiner Konzeption des Imaginären selbst: das instituierte Imaginäre kann nur von einem anderen, neuen Imaginären infragegestellt werden, einer neuen 'Idee' — was den Charakter des Imaginären als unmittelbare Produktion bestätigt. Aber genau hier liegt die Gefahr, das Imaginäre durch diese Heiligung zu ontologisieren, indem es derart verabsolutiert wird: es ist eine Sache zu sehen, daß das Symbolische einer Gesellschaft, das — Castoriadis: — institutionalisierte Imaginäre von einer 'Idee' getragen ist, die eine *creatio ex nihilo* des 'Ideenhabers' sein mag, und eine andere zu erklären, warum sich denn diese und keine andere durchsetzt, gesellschaftlich Geltung erlangt, und überhaupt instituiert werden kann, das heißt: Funktion übernehmen kann. Genau dann aber verliert das Imaginäre seine unmotivierte Gestalt und wird zu einer gesellschaftlichen Funktion, die alles andere als willkürlich oder unmittelbare Produktion ist, sondern vielmehr vermittelte und vermittelnde gesellschaftliche Dimension in Bezug auf, also relativ zu und nicht absolut von, anderen Bereichen. Castoriadis vollzieht *de facto* genau das, was er de verbo den Strukturalisten vorwirft: eine Entgeschichtlichung der Gesellschaft; denn betrachtet man — wie er — nur die Ordnungen als Ordnungen, also synchron, so sind sie ebenso willkürlich wie die Mode, nur eben langlebiger. Sie gewinnen erst 'Sinn' als 'Funktion', das heißt in Bezug auf andere Ordnungen. Castoriadis kann das nicht sehen — jedenfalls nicht systematisch in seine Theorie einbauen — weil er diese organisierende Ordnung der Gesellschaft — das muß nicht die Ökonomie sein — selbst als institutionalisiertes Imaginäres statt als reale gesellschaftliche Struktur begreift; Herrschaft als Grundstruktur einer Gesellschaft wird somit ein Instituiertes, anstatt ihrer Logik entsprechend ein Instituierendes.

Dogmatisch gesprochen formuliert Castoriadis damit einen Überbauprimat:

> die zentralen oder primären Bedeutungen einer Gesellschaft *schöpfen* nämlich Objekte *ex nihilo* und organisieren die Welt (als Welt 'außerhalb' der Gesellschaft, als gesellschaftliche Welt und deren wechselseitige Verschränkung) (...) Die zentralen Bedeutungen sind nicht Bedeutungen 'von' etwas — und allenfalls in einem abgeleiteten Sinne an etwas 'gebunden' oder auf etwas 'bezogen'. Sie sind das, was die scheinbar verschiedenartigsten Gegenstände, Handlungen und Individuen einer bestimmten Gesellschaft zugehören läßt. Sie besitzen keinen 'Referenten'; sie instituieren eine Seinsart von Dingen und Individuen, und zwar so, daß diese Dinge und Individuen auf jene Bedeutungen bezogen sind[59];

Castoriadis führt als ein Beispiel solcher zentraler gesellschaftlich imaginärer Bedeutungen 'Gott'' an[60].

Das Problem dieser Konzeption des Imaginären ist weniger, daß Castoriadis damit dem 'Überbau' einen logischen Primat einräumt, denn in der Tat fällt es schwer, das Symbolische eindeutig als abgeleiteten Überbau oder als produzierende Basis festzulegen, zumal die Konzeption des Symbolischen wohl eher eine ist, die diese rigide Dichotomie aufheben soll; das Problem ist vielmehr, daß diese *creatio* schwer begreiflich ist als eine *ex nihilo*, da nicht zu sehen ist, wie sie sich ohne Funktionsbezug erhalten soll, das heißt 'Anwendung' und, gesellschaftliche Akzeptanz finden, ja überhaupt entstehen kann. Was Castoriadis hier undeutlich macht, ist die Frage nach der Spezifität einer solchen Vorstellung in einer bestimmten Gesellschaft, also die Frage danach, warum gerade diese Vorstellung in diesem Netz von Bedeutungen in dieser Gesellschaft zentral oder überhaupt ist. Wenn er den Zusammenhang umgekehrt deutet: daß diese zentrale imaginäre Bedeutung die Form der Gesellschaft ermöglicht, so fällt er hinter Hegel zurück, erst recht hinter Marx; wenn er die Beziehung als Funktionszusammenhang deuten würde, müßte er das 'ex nihilo' aufgeben, also seine gesamte Konstruktion des Imaginären, da eine Funktionalität *ex nihilo* nicht vorstellbar ist; beide Fälle geraten auf die eine oder auf die andere Weise mit den Intentionen seiner Gesellschaftstheorie in Konflikt.

Ein weiteres theoretisches Problem, das hierbei auftritt, ist, daß nicht mehr erklärt werden kann, warum bestimmte zentrale Bedeutungen imaginär im 'üblichen' Sinn, also ideologisch sind, wenn es so sein soll, daß "die Bedeutungen nicht *das* sind, was sich die Einzelnen bewußt oder unbewußt vorstellen oder was sie denken. Sie sind vielmehr das, wodurch und von wo aus die Individuen als gesellschaftliche Individuen formiert, das heißt befähigt werden, am gesellschaftlichen Tun und Vorstellen-Sagen teilzunehmen"[61]. Sie sind somit unmittelbar 'real', denn — in Anspielung auf eine marxistische Interpretation von Lacans imaginärer Repräsentation oder des Symbolischen —:

> wir dürfen uns die Welt der gesellschaftlichen Bedeutungen also weder als ein irreales Doppel einer realen Welt denken noch als anderen Namen für ein hierarchisches System von 'Begriffen' (...) Wir haben die Welt der gesellschaftlichen Bedeutungen als ursprüngliche, anfängliche und irreduzible Setzung des Gesellschaftlich-Geschichtlichen und des gesellschaftlichen Imaginären zu denken, so wie es sich in einer Gesellschaft jeweils zeigt (...) Es ist diese Institution von Bedeutungen, die – gestützt auf die Institution des *legein* und *teukein* – festlegt, was für eine Gesellschaft ist und was nicht, was Wert hat und was wertlos ist, aber auch *wie* dasjenige, dem Sein oder Wert zukommen kann, ist oder nicht ist[62].

Sie bestimmen also die Form und den Inhalt der gesellschaftlichen 'Realität', nicht diese die Bedeutungen.

Castoriadis denkt das *Imaginäre* somit eindeutig als *Imagination*, als Prozeß und Akt der Sinnsetzung im schöpferischen Sinne: als Entwicklung von Bezeichnungen, die die Gesellschaft strukturieren; dafür wird in der Regel der Begriff des Symbolischen verwendet, der diese 'Kreativität' genauso enthält, aber als mit der Gesellschaftsstruktur vermittelter. Das Imaginäre als Imagination zu denken, macht es, wie bei Castoriadis, der absoluten Schöpfung, dem Ende der gesellschaftlichen Funktion als Vermittlung gleich und läßt die analytische Schärfe des Begriffs untergehen; das Imaginäre muß als funktionaler gesellschaftlicher Bereich, nicht als reine Aktivität gedacht werden, denn die Tatsache, daß irgendetwas Setzung und Schöpfung ist, ist letztlich trivial, weil alles — jeder Satz, jede Handlung, jeder Gedanke — in diesem Sinn 'Schöpfung' ist; der Begriff sagt dann nichts mehr aus.

Genaugenommen sind Castoriadis' Begriffe so allgemein gefaßt, daß sie für alle Gesellschaften zutreffen, auch für eine 'ideale', die eben auch eine instituierte Ordnung hätte, eine imaginäre, das heißt geschöpfte, selbstgesetzte 'Logik'; die Begriffe in Castoriadis' Theorie der Gesellschaft gewinnen daher einen ontologischen Charakter und verlieren dadurch die analytische Schärfe der Bestimmtheit, weil sie ihre normative Kraft einbüßen: ein Begriff wie der des Imaginären, der nicht gekoppelt wird an den der Herrschaft, sagt eben nicht mehr aus als der Begriff des Symbolischen, der die allgemeine Tätigkeit des Menschen festhält, Zeichen zu bilden, Welt zu deuten, Deutungsmuster zu konstituieren und das heißt: Gesellschaft zu 'konstruieren'. Der bestimmte 'Inhalt' oder die Form dieser Symbolsysteme — ob sie nun imaginär sind oder nicht — kann nicht erfaßt werden; eigentlich kann Castoriadis den entscheidenden 'Akt' gesellschaftlicher Praxis, der in Wirklichkeit eine Struktur ist, nämlich die Vermittlung, gar nicht denken, denn "die Gesellschaft ist keine Menge, kein System und keine Hierarchie von Mengen oder Strukturen (die sie als Organisation im Sinn gesellschaftlicher Vermittlung denken ließe — FL); sie ist Magma und Magma von Magma"[63]. Die Gesellschaft als selbstbezügliches Magma und die identitätslogische Organisation dieser Gesellschaft ist nur eine Dimension dieses Magmas, denn "gewiß wäre es ein entscheidender Irrtum, ein Mord am Objekt — die Untat des Strukturalismus! — wenn man annehmen wollte, in dieser Logik gehe das Leben oder auch nur die Logik einer Gesellschaft auf"[64]. Hier ist die Stelle in Castoriadis' Gesellschaftstheorie, wo sie von einer möglicherweise fruchtbaren methodischen Kritik am Strukturalismus[65] umschlägt in eine *lebensphilosophische Konstruktion des Sozialen*; statt einer Hinwendung zur Organisation der Lebenswelt der Gesellschaft erfolgt die unterschwellige Rückbesinnung auf einen dionysischen

Nietzsche, die Castoriadis in merkwürdige Nähe zum Poststrukturalismus der Marke Deleuze/Guattari bringt: Magma, Ströme, Pulsieren, Gesellschaftstriebe, gesellschaftlicher Wille des Imaginären zur Macht. Auf die Dauer verliert dabei der Begriff 'Gesellschaft' jede Bedeutung und wird zu der von 'Leben' überhaupt, ein Wort, das Castoriadis, wie im obigen Zitat, des öfteren verwendet; bedenklich ist die Naivität, mit der Castoriadis ähnlich wie die Jugendbewegten mit den besten Absichten diese lebensphilosophischen Begriffe verwendet: seine Gesellschaftstheorie setzt sich so — gegen jede Intention seinerseits — der Gefahr aus, integrierbar zu werden ins fröhlich imaginäre posthistoire.

Um dies zu verhindern, sollte das Symbolische weiterhin als Funktion des Realen in einer Gesellschaft gesehen werden, da es erst von dort her seinen Sinn erhält; ist diese Gesellschaft durch Herrschaft strukturiert, so ist das Symbolische in seinen gesellschaftlich entscheidenden Bereichen imaginär im Sinn von ideologisch als Legitimation dieser Herrschaft; das Imaginäre verliert bei Castoriadis diese kritische Bedeutung, da er ihm den normativen Gehalt dadurch entzieht, daß er es zum Motor der Geschichte verklärt.

Castoriadis, der sich von Lacans Begriff des Imaginären absetzen muß, um seinen zu konzipieren, verkennt durch seine vorschnelle Kritik an Lacans Konzeption deren kritische Dimension, indem er Lacans versteckte Ontologie überbetont anstatt seinen Ansatz für eine kritische Konzeption des Imaginären fruchtbar zu machen: "das Spekulare, 'Spiegelhafte', das offensichtlich nur ein Bild *von*, ein reflektiertes Bild ist, anders gesagt: das *Widerspiegelung* und damit ein Abfallprodukt der platonischen Ontologie (des *eidolon*) ist"[65]. Durch seine Idiosynkrasie gegenüber Lacan verschenkt Castoriadis die kritische Dimension des Begriffs des Imaginären, wenn er überzogen polemisch — wie leider so oft in seinem Buch und nicht nur gegen Lacan — formuliert: "tatsächlich gewinnt Lacans Begriff des 'Imaginären' nur dann einen Anschein von Sinn, wenn es auf ein Reales bezogen und einer Realität gegenübergestellt wird, *die sich mit Händen greifen läßt*"[66]; Castoriadis hat nicht begriffen, oder besser: kann mit seinem Begriff des Imaginären nicht begreifen, daß das Imaginäre bei Lacan nicht etwa, wie er zu glauben scheint, irreale Illusion, Halluzination, sondern vielmehr und ein Moment der Realität selbst ist.

Castoriadis macht einen — weil auf diesem Ansatz seine ganze Analyse gründet — fundamentalen Fehler: 'imaginär' in Castoriadis'

Bedeutung als Begriff der Kritik bedeutet nicht etwa, wie bei Lacan, die mangelhafte Einsicht in die Realität gesellschaftlicher Bezüge, sondern die Unangepaßtheit an, die Ungleichzeitigkeit zur gesellschaftlichen Struktur. Castoriadis stellt für die heutige Subjektverfassung das Fehlen einer "integrierten Persönlichkeitsstruktur"[67] fest und interpretiert dies als Versagen der Gesellschaft, die ihre Individuen nicht mehr integrieren kann beziehungsweise der sich ihre Individuen nicht mehr integrieren können, wie dies bei vorherigen, zum Beispiel archaischen Gesellschaften, deren Individuen nicht weniger 'neurotisch' gewesen seien als die heutigen — nur daß eben deren Neurose der Gesellschaftsform konform gewesen sei — der Fall gewesen sei. Die gegenwärtige psychische Verfassung der Subjekte — und es bleibt etwas unklar, ob er dabei die 'Neurotiker' im klinischen, also manifesten Sinn meint oder im latenten als prinzipielle Psychostruktur — erscheint bei Castoriadis als für die Gesellschaft dysfunktional, da sie ihr nicht mehr 'entspricht'; woher sein Kriterium für Angemessenheit, Entsprechung stammt, wird von Castoriadis nicht erläutert, er selbst zeigt keines auf, führt nur den Begriff funktionaler Entsprechung ein, ohne diese Funktion näher zu charakterisieren. Für Castoriadis ist somit klar erwiesen,

> daß der Persönlichkeitstyp des Spartaners oder Mundugumor trotz möglicherweise neurotischer Anteile seiner Gesellschaft *funktional entsprach*, daß sich der Einzelne mit seiner Gesellschaft in Übereinstimmung fühlte, daß er sie seinen Ansprüchen gemäß funktionieren lassen und die nächste Generation dazu heranbilden konnte. Die 'Neurose(n)' der heutigen Menschen stellen sich, soziologisch betrachtet, hingegen meist als Erscheinungen der Fehlanpassung dar und werden nicht nur subjektiv als Unglück erlebt, sondern hemmen vor allem auch das gesellschaftliche Funktionieren der Individuen. So können diese nicht mehr angemessen auf die Anforderungen des Lebens, wie es ist, reagieren, und in der nächsten Generation reproduziert sich die Unangepaßtheit auf erweiterter Stufenleiter. *Die 'Neurose' des Spartaners ermöglichte ihm gerade seine Integration in die Gesellschaft, während der moderne Mensch von seiner 'Neurose' daran gehindert wird* (...denn – FL) die Bildungen des Unbewußten entsprechen den Regeln im soziologischen Sinne nicht mehr[68].

Castoriadis verkennt hier, daß gerade diese nicht vorhandene, also 'neurotische' Persönlichkeitsstruktur diese unsere Gesellschaft am Laufen hält und extrem funktional ist, da sie die subjektive Vermittlung der objektiven Struktur ist; und er muß dies deswegen verkennen, weil er das Verhältnis von Individuum und Gesellschaft falsch denkt, nämlich als Partizipation statt als Funktionalität. Da Castoriadis überall der Theorie des Subjekts als Funktion im gesellschaftlichen Feld entgegentreten will, muß er das Imaginäre dieser Beteiligung verkennen, denn diese Integration durch pseudohafte Beteiligung ist gerade der vom ideologischen Imaginären erzeugte Schein, der die Funktionalisierung und Verdinglichung der Subjekte verklären soll; denn die 'Bildungen des Unbewußten', die Struktur des Unbewußten entspricht funktional der Gesellschaftsstruktur, deren Realität wahrzunehmen verhindert wird. Solange die Gesellschaft selbst imaginär strukturiert ist durch Herrschaft, muß — für deren Akzeptanz — auch die Psyche der Individuen imaginär organisiert sein; wenn Castoriadis 'imaginär' als Fehlanpassung an die von der jeweiligen Gesellschaft ausgegebene Realität begreift, so unterschlägt er damit, daß diese — den Subjekten durch Sozialisation vermittelte — 'Realität' imaginär sein kann. Sein Begriff des Imaginären bleibt als kritischer an der gesellschaftlich organisierten, gedeuteten, angebotenen 'Realität' kleben, ohne diese selbst infragestellen zu können, weil ihm der normative Kontext verloren gegangen ist.

Castoriadis kann nicht — wie etwa Baudrillard — sehen, daß sich im Verschwinden der positiven, nicht einfach nur negativen Ordnungen[69]: im bekannten Verfall allgemein gültiger Werte, ein Wandel in der Struktur des Ideologischen, also in der Sphäre der Legitimation, manifestiert: von der inhaltlich bindenden, in gewisser Hinsicht 'argumentierenden' Ideologie zur formalen Kulturindustrie, wodurch dem System enorme legitimatorische Freiräume für technokratisches Handeln gegeben werden. Castoriadis verkennt, daß die verdinglichten Individuen dafür funktional, ja sogar notwendig sind, da Massenloyalität bereitgestellt werden muß; sein Begriff des Imaginären kann keine 'Realität' bezeichnen, denn Castoriadis begreift die geschilderten Tatsachen als "Machtverlust des instituierten Imaginären"[70], anstatt sie als Perfektionierung der Herrschaft durchs Imaginäre erkennen zu können, weil er unter dem Imaginären ein unmittelbar Reales versteht statt ein als Funktion des Realen Vermitteltes. Somit trennen sich für Castoriadis heute Weltbild und Gesellschaftsbild, die bisher stets synchron, das heißt von der Struktur her homolog waren, und verursachen dadurch die "Krise des (instituierten) Imaginären in der modernen Welt"[71], ohne daß er erkennen könnte, daß eben diese Subjektentmachtung — die er als Krise des Imaginären begreift, das das Subjektive repräsentiert — und deren Kompensation eine zentrale Funktion des ideologischen Imaginären ist.

Entsprechend ist dann auch die Kritik, die Castoriadis am kritischen Gebrauch des Begriffs 'imaginär' übt, der das Imaginäre als Kompensation eines realen Mangels begreift. Castoriadis greift diese Konzeption des Imaginären als "Kompensation eines Defizits" dahingehend an, daß er ihre Intention verkehrt: das Defizit erscheine nur auf der Folie abendländischer Wissenschaft und Rationalität im europäischen Sinn und somit "die Vorstellungen der Wilden als der Versuch, die Löcher zu stopfen, die sie in der Organisation *ihrer* Welt *hätten finden müssen, wenn sie von seinen* (des europäischen Wissenschaftlers — FL) Phantasmen beherrscht gewesen wären"[72]. In Wirklichkeit beruht der Ansatz der Auffassung des Imaginären als Kompensation allerdings weniger auf Ethnozentrismus beziehungsweise auf der Übertragung intellektueller Phantasien auf andere soziale Schichten, sondern auf der Erkenntnis gesellschaftlicher Mängel — also letztlich Herrschaft — die im Imaginären kompensiert werden; das ist ja eine der grundlegenden Intuitionen der Aufklärung überhaupt, die sich in der Romantik noch vorwiegend ästhetisch organisiert, indem die Kunst, die Poesie als Kompensation, als einheitsstiftende Utopie der Versöhnung der Momente einer entfremdeten, zerfallenen Realität auftritt, was als Aufgabe im deutschen Idealismus vor allem der Vernunft als Philosophie zufällt. Habermas hat zurecht eine Parallele zwischen der selbstgestellten Aufgabe der hegelschen Philosophie und der gesellschaftlichen Funktion des Mythos gezogen[73]; diese Intuition der Aufklärung findet dann ihre ausdrücklich kritische Fundierung — vom jungen Hegel schon ansatzweise vorgedacht[74] — in Feuerbachs Religionskritik, die sich beim jungen Marx fortsetzt, als Ideologiekritik paradigmatisch wird und in den Analysen des Warenfetischs und kapitalistischen Scheins die Verschiebung zur Kulturindustrie, ohne sie als solche begreifen zu können, schon geahnt hat; all dies verschwindet bei Castoriadis unter dem Teppich des Vorwurfs der Kulturblindheit. Anstatt diesen Zusammenhang zu sehen und den 'realen' Charakter des Imaginären im kritischen Sinn zu erkennen, vergibt Castoriadis diese Chance, indem er es — wie in seiner Abgrenzung von Lacan schon gesehen — auf Illusion und Irrealität reduziert.

Aber die 'Seinsweise' des Imaginären ist nicht Nichtsein, sondern Schein: die durchs Imaginäre vermittelte Beziehung zwischen der Realität und dem Subjekt ist die Suggestion, Imaginierung, Vorspiegelung von Realität, einer imaginären Konstruktion der Wirklichkeit; die Aufgabe des Imaginären ist die der *Ästhetisierung der Realität*. Die Zeichen im Imaginären sind nicht kommunikative Symbole, sondern

festgefrorene Einheiten, pathologische Zeichen, die für sich selbst
stehen[75]. Und wo es Castoriadis einmal gelingt, diese heutige Struktur
des Imaginären als Enteignung der kommunikativen Qualität des
Symbolischen zu fassen, da muß er letztlich diese Möglichkeit
unrealisiert vorüberziehen lassen, da er mit seinem Begriffsapparat die
folgende überzeugende Aussage, die ihn in die Nähe Baudrillards und
seiner überzeugenderen Theorie des Imaginären hätte führen können
verschenken muß "das *charakteristische* und tiefgreifendste, folgen-
reichste, aber auch vielversprechendste Merkmal des modernen
Imaginären... dieses Imaginäre hat *kein eigenes Fleisch*, es erborgt seine
Materie immer etwas anderem. Es ist phantasmatische Besetzung,
Aufwertung und Verselbständigung von Elementen, die als solche nicht
zum Imaginären rechnen, sondern zum — begrenzten — Rationalen
des Verstandes und des Symbolischen"[76]. Dies bleibt unbegriffen da
er prinzipiell "die Vorstellung, die Imagination, das gesellschaftliche
Imaginäre"[77] als Phantasie begreift statt als den Tod der Phantasie
durch die Fixierung des Symbolischen als Wahrnehmungsverhinderung.

Das Imaginäre ist das Wesen, die Struktur der Verdinglichung; es
bezeichnet den Charakter eines Prozesses und Zustandes der kapitalistisch organisierten Gesellschaft, der zu ihrer Aufrechterhaltung
notwendig ist; es muß daher als Prozeß und Struktur zugleich begriffen
werden. Ein sinnvoller Begriff des Imaginären muß die beiden Ebenen
der genetischen und logischen, diachronen und synchronen, historischen und strukturellen Dimension der Gesellschaft umfassen und in
Bezug setzen können; es ist ein Begriff, der Strukturalismus und
Kritische Theorie verbindet in der gesellschaftstheoretischen Betrachtung dessen, was das Imaginäre als gesellschaftlichen Bereich in der
Realität und im Bewußtsein der Individuen ausmacht: die Struktur
formaler Organisation von Institutionen[78] und von Bewußtsein — der
Verdinglichung.

Imaginäres in diesem Sinn gibt es nur, weil und solange es
Herrschaft als gesellschaftliche Struktur gibt, als deren reale, das heißt
real wirksame Legitimation und Massenloyalität sichernde Bewußtseins-
struktur es fungiert; es kann nicht von dieser Funktion abgekoppelt
werden, wie Castoriadis das in seinem Entwurf des gesellschaftlichen
Imaginären vorzuführen versucht, da es immer ein Imaginäres gibt, das
diese legitimatorische Funktion wahrnimmt, selbst wenn es zwischen
den Inhalten dieses Imaginären scheinbare Widersprüche gibt, wie zum
Beispiel zwischen den Forderungen nach christlicher Moral und

Nächstenliebe gegenüber der nach ökonomischer Konkurrenz. Das Imaginäre ist real dadurch, daß es in imaginärer Form die symbolische Funktion, die 'Ordnung des Diskurses' konstituiert, indem es das — in der kommunikativen Potenz der Sprache enthaltene — 'emphatische' Moment dieses Symbolischen: die Möglichkeit selbstorganisierter Verständigung, abgespalten hat und zugleich die so entstehende Lücke mit anderen Formen aus dem imaginären Bereich der Gesellschaft: vor allem durch Kulturindustrie, kompensiert und durch Identifikationen scheinhaft schließt.

Was für Castoriadis als imaginäre "Vorstellungen ... als das konkrete Besondere *par excellence*"[79] erscheint, also als Ausdruck individueller Totalität, beruht als gesellschaftliches Imaginäres strukturell darauf, daß ein Individuum nur so phantasieren, vorstellen und kompensieren kann, wie ihm dazu gesellschaftliche Signifikanten zur Verfügung stehen; wenn das Symbolische durchs Imaginäre enteignet ist, wird Castoriadis' radikale Imagination zum Ritual der Wiederholung; demgegenüber erscheint seine 'freie Imagination' als bloßes Postulat einer Sehnsucht.

Imaginäre Sozialisation

So gerät bei Castoriadis das 'radikale Imaginäre' bei dem ontologischen Status, den er ihm zuweist, von vornherein als nahezu identisch mit dem Begriff des 'Lebens', wie ihn die Lebensphilosophie kennt, oder vielmehr mit dessen Motor, dem *elan vital*. Dasselbe lebensphilosophische Begreifen der Gesellschaft, der Welt nach dem Modell der 'Seele' vollzieht auch Castoriadis: seine Gesellschaftstheorie mit den Gesellschaftstrieben des Imaginären als Zentrum ist eine Verseelung der Gesellschaft, in der als Modell für das Imaginäre das Es herhalten muß, dessen Magma plötzlich triebhaft aus dem Unbewußten ausbricht und sich im Bewußtsein instituiert; gesellschaftliche Prozesse, der Prozeß der Imagination und der Instituierung des Imaginären, genauer: des Imaginierten werden psychisch, wie in der individuellen Psyche begriffen, analog zu der Weise, in der der Psychoanalytiker Castoriadis das Psychische begreift.

Castoriadis sieht die strukturelle Interpretation der Psyche, wie sie Z. B. von Lacan geleistet wird, als theoretische Verdinglichung an; daß man "aus dem Psychischen eine reale Maschinerie gemacht beziehungsweise es auf eine logische Struktur herabgesetzt hat"[80], erscheint ihm als der Hauptmangel der lacanschen Theorie. Stattdessen besinnt er sich wie üblich zurück aufs radikale Imaginäre und setzt dieses der verdinglichenden Strukturierung des Psychismus entgegen, nämlich das "Sein der Psyche, die nichts anderes *ist* als Entstehung von Vorstellungen", also "radikal anders, als alles, was uns im Rahmen irgendeiner Sprache oder Algebra begegnen könnte"[81]; somit existiert für Castoriadis "die Psyche als radikale Imagination — das heißt im wesentlichen als Auftauchen von Vorstellungen, als ein Vorstellungsstrom, der der Bestimmtheit nicht unterworfen ist"[82]. Man muß sich dabei vor Augen halten, daß dies ein ausgebildeter Psychoanalytiker schreibt, der selber Patienten behandelt und im Vorwort behauptet, dem Studium der psychoanalytischen Theorie "den größen Teil der Jahre 1965 bis 1968 gewidmet"[83] zu haben; so, als ob es noch nie erfolgreiche Versuche gegeben hätte, die Psychoanalyse gesellschaftstheoretisch zu begreifen, schreibt Castoriadis dem Psychischen — in Analogie zum Gesellschaftlichen — wiederum eine absolute, eigenständige Existenz zu, die ihr als Psyche gegeben sei, sozusagen als ontologische Qualität, denn "schon der 'allererste' Schritt bei der Konstitution dieser Erfahrung (der der Ver- und Bearbeitung von 'Eindrücken' aus dem Realen - FL) setzt voraus, daß die Psyche als — sei's auch noch so rudimentäre — *Erfahrung* zu organisieren vermag, was sonst Chaos innerer und äußerer Eindrücke bleibe (...). Die Psyche ist ein *Formant*, der nur in dem von ihm Geformten, durch es und *als* dieses besteht. Sie ist Bildung (i. O. deutsch) (*formation*) und Einbildung (i. O. deutsch) (*imagination*), radikale Imagination, die aus einem Nichts an Vorstellung, das heißt *aus nichts* eine 'ursprüngliche' Vorstellung auftauchen läßt"[84]. Sie ist "ursprüngliches Vermögen, Vorstellungen auftauchen zu lassen"; Castoriadis meint damit, ist, daß vor aller Tätigkeit, vor aller psychischen Aktivität ein Grund da sein muß, auf dem alles weitere aufbaut, eine erste Vorstellung, denn "vor allem ist 'zu Beginn' eine 'erste' Vorstellung erfordert, die die Möglichkeit zur Organisation aller anderen Vorstellungen gewissermaßen in sich schließen muß. Diese erste Vorstellung wäre ein Formend-Geformtes, eine Figur, die alle Schemata der Figuration keimhaft enthalten müßte und in der, wie embryonal auch immer, alle Organisationsschemata angelegt sein müßen, aus denen sich dann die psychische Welt entwickelt"[85]. Diese transzendentale 'erste Vorstellung' als Bedingung der Möglichkeit von psychischer Vorstellung überhaupt ist allerdings deshalb 'erfordert', weil Castoriadis mit seiner Theorie des radikalen Imaginären es anders gar nicht denken kann; ohne diese erste Vorstellung oder das radikale Imaginäre als Wurzel aller Imagination erschiene das gesellschaftliche Imaginäre nämlisch als 'Resultat' — um mit Hegel zu sprechen — statt als Urgrund. Castoriadis erweist sich so als ein daseinsanalytischer Kantianer mit transzendentaler Seelenlehre: eine keimhafte erste Vorstellung als Bedingung der Möglichkeit aller anderen, indem sie schon keimhaft, embryonal die Schemata der Weltkonstitution enthält — die sie doch noch gar nicht kennen kann, da es 'ursprünglich' weder eine Welt noch eine Psyche gibt und Psyche erst entsteht im Vollzug der Weltaneignung; selbst wenn Castoriadis seine ontologischen und anthropologischen Wörter in Anführungszeichen setzt, ändert das nichts: er kann die Intersubjektivität als Prozeß und Struktur der Entwicklung der Psyche und der Konstitution von Welt nicht wirklich denken, bleibt beim individuell verankerten 'Vermögen' hängen und somit bei Kant, vor Hegel.

Jeder Psychoanalytiker weiß doch, daß die Psyche "durch Erfahrung erworbenes Bewußtsein"[86] ist und nicht das Ergebnis eines radikalen Imaginären; diesen Prozeß der Erfahrung als einen intersubjektiven kann Castoriadis gar nicht denken durch seine Ontologisierung des Imaginären, die er anthropologisch darin verankert, "daß die ursprüngliche Phantasiebildung — das, was ich radikale Imagination nenne — noch der primitivsten Trieborganisation vorausgeht und noch vor dieser existiert"[87], daß also der Mensch anthropologisch mit dem Imaginären als Vermögen ausgestattet ist. Es ist aber — bei einem intersubjektiven Ansatz — gar nicht nötig, diese Wesensannahme zu machen, da die 'ursprüngliche Phantasiebildung' 'vor aller Trieborganisation' auf die privilegierte, ja vorerst einzige Lebenserfahrung des Kleinkinds zurückführbar ist: es gibt keine 'ursprüngliche Phantasiebildung vor den Trieben', weil es immer — selbst noch vor der Geburt, wo es ja im Normalfall wegen nicht vorhandenem Defizit keinen Grund zur imaginären Komponsation gibt — den Körper des Kindes gibt, und Defizite werden in den entscheidenden Monaten des Säuglingslebens primär dort erfahren. Aber Castoriadis kann diese 'materialistische' Basis der Psyche nicht begreifen, denn er versteht den erfahrenen Mangel des Säuglings, der die Phantasie und das Imaginäre als Kompensation — wie es besonders deutlich wird beim 'Übergangsobjekt' — produziert, als selbst vom Imaginären produziert, da ihm eine Erwartung, ein Vorstellen der Fülle vorausgehen müsse[88]; dieser

Mangel aber wird nicht geistig erlebt, sondern körperlich im Kontakt mit der Mutter, also intersubjektiv erfahren: da es nämlich in der Schwangerschaft die reale Erfahrung der Fülle gab und da es somit eine Körperorganisation gibt, die gewöhnt ist, versorgt zu werden, ihre 'Identität' zunächst nur finden kann in der totalen Versorgung. Die 'Vorstellung' der Fülle ist kein geistiger Akt, kein Bewußtsein von Möglichkeit, sondern eine körperliche 'Konditionierung' durch eine zuvor perfekte und permanente Versorgung; der Mangel durch die reale Entbehrung nach der Geburt ist körperliche Unlust, die — bei Vorhandensein der berühmten 'genügend guten Mutter', also im intersubjektiven Kontakt — durch Phantasie kompensiert werden kann.

Diese Fehleinschätzung von Castoriadis resultiert aus dem Blickwinkel der 'psychischen Monade', den er vertritt, indem er behauptet, daß diese ursprünglich vorhanden sei und erst später sozialisiert werde in "der gesellschaftlich-geschichtlichen Institution des Individuums (sowie, im Zusammenhang damit, der Wahrnehmung und des Dings). Gemeint ist die Verwandlung der psychischen Monade in ein gesellschaftliches Individuum, für das es andere Individuen, Objekte, eine Welt, eine Gesellschaft und Institutionen gibt — Dinge also, die für die Psyche ursprünglich weder Sinn noch Sein haben"[89]. Aber außer dem, was in dieser Reibung, in dieser Intersubjektivität entsteht, wäre doch sonst nichts: die Monade ist nicht nur fenster-, sondern vor allem geist- und seelenlos; somit gibt es auch keine Verwandlung, kein gesellschaftliches Zauberkunststück oder keine sozialisatorische Eucharistie — Individuen sind immer schon gesellschaftlich, in Gesellschaft, intersubjektiv in Kontakt, so erst werden sie Subjekte und es gibt kein 'Vorher'.

Castoriadis scheint bei der Untersuchung der Psyche das, was man 'psychische Realität' nennt, für die Sache selbst zu nehmen, anstatt zu sehen, daß diese einer bestimmten Struktur folgt und eine bestimmte Struktur von Übertragungen, Wiederholungen, Identifikationen und nicht etwa "das Sein der Psyche"[90] ist. Aber genau diese Sichtweise kann Castoriadis nicht vertreten, will er seine Konzeption des Imaginären retten, das als Quell aller Bedeutungen auch der Psyche zugrundeliegt und diese 'motiviert' und nicht etwa als Unbewußtes — wie Lacan behaupten würde — selber strukturiert ist: "das Unbewußte besteht nur als ungeteilter Strom von Vorstellungen (Affekten), Intentionen"[91]; da kommt dann wieder diese unselige Metapher zu ihrem (Un) Recht: unter der Oberfläche brodelt das Magma, ein Strom von Materie ohne Ordnung, 'ungeteilt' und undifferenziert. Castoriadis

macht hier offenbar den in Frankreich — vergleiche Deleuze/Guattari
— recht beliebten Fahler, Unbewußtes und Es miteinander zu
identifizieren; mag das Es noch so magmatisch sein, es tritt dennoch
niemals auf, ohne die Organisation des Unbewußten zu durchqueren
und somit von ihr strukturiert zu werden, denn selbstredend besitzt das
Unbewußte eine Struktur, und zwar eine, die es mit dem Ich verbindet,
dessen größter Teil bekanntlich ebenfalls unbewußt ist. Wenn Freud
schreibt, es gebe 'dort' keine Zeit, keine Verneinung und keine
logischen Widersprüche, so besagt das nur, daß die Struktur des
Unbewußten eben eine ist, die chronologisch und logisch nahezu alles
zu integrieren vermag, da das Unbewußte diese Grenzen nicht kennt,
also zu Wiederholungen neigt, genauer: daraus besteht; es ist 'sich'
nicht 'radikale Imagination', sondern Struktur.

Castoriadis muß diese Ontologie der Psyche konzipieren, will er
nicht sein Imaginäres desavouieren; die Psyche ist aber nicht ein von
der Gesellschaft an sich entfremdetes Wesen, eine Monade, wie
Castoriadis das sieht, sondern sie ist überhaupt nur als gesellschaftliche
vorstellbar. Das Unbewußtes ist die Instituierung der Gesellschaft im
Subjekt, die Castoriadis nur deshalb notwendig als Joch erscheint, weil
er einen Naturzustand konstruiert, den es nie gab: "das gesellschaftliche
Individuum ist, so wie die Gesellschaft es erzeugt, nicht 'ohne
Unbewußtes' vorstellbar; die Institution der Gesellschaft, mit der
untrennbar die Institution des gesellschaftlichen Individuums verbunden ist, zwingt der Psyche eine Organisation auf, die ihr wesenhaft
fremd ist"[92]. Wesenhaft fremd muß dies in Castoriadis' Theorie erscheinen, weil er das Wesen als Magma begreift, das als Monade mit der
Gesellschaft konfrontiert wird, statt überhaupt erst im Kontakt mit
dieser zur Existenz zu kommen. Castoriadis stilisiert damit die
narzißtische Problematik zu der des aufwachsenden Kindes überhaupt:
ontologisch erscheint die schrittweise Aneignung der Realität durch das
Kleinkind als dessen Übermächtigung durch die Gesellschaft; die
Sozialisation müßte nach dem Modell von Castoriadis als solche zu
narzißtischen Störungen führen, da er sie nur auf der Folie einer
rigiden Entgegensetzung des monadischen, hilflosen Subjekts und der
übermächtigen Gesellschaft konzipieren kann — daß das eine typisch
kleinbürgerliche, wenn auch auf der Oberfläche nicht unwahre
Wahrnehmung: Gesellschaft versus Ich ist, sei hier nur am Rande
vermerkt — ohne das Individuum in einem intersubjektiven Bezie-
hungsfeld zu sehen, *dessen* Struktur darüber entscheidet, ob das Kind

der Realität brutal und entsprechend hilflos ausgeliefert wird oder
durch die familiäre Fürsorge davor werden bewahrt kann.

Was bei Castoriadis verlorengeht, ist die mögliche Differenzierung,
daß es von der jeweiligen Struktur einer Gesellschaft – und damit auch
ihrer Sozialisation und deren gesellschaftlichem Raum in Familie und
Schule etcetera – abhängt, ob die Sozialisation eine Vergewaltigung
darstellt, weil sie auf die Verdinglichung der Subjekte hinausläuft, oder
nicht; ob es also eine generelle Aneignung des Subjektiven durch die
Gesellschaft, also die Enteignung des Subjekts im Prozeß seiner
gesellschaftlichen 'Herstellung' gibt, oder ob es Gesellschaftsstrukturen
geben kann, die die Subjektivität erst richtig zu sich kommen lassen.
Castoriadis kann durch seine Aufrichtung der Statue des radikalen
Imaginären als anthropologischer Konstante die Sozialisation immer
nur als Entfremdung denken, da er einen grundsätzlichen Gegensatz
von Gesellschaft und Individuum konstruieren muß, um die Legitimation seiner Monadologie und damit des Imaginären als Magma zu
retten; die Intersubjektivität als Konstitution von Subjekten bleibt bei
Castoriadis außen vor. Gesellschaft und Subjek bleiben somit immer
Gegner: Kommunikation im Sinn von Verständigung kann eigentlich
nicht gedacht werden.

Ontologie des Imaginären

Castoriadis hat somit eine grundsätzliche und, wie noch zu sehen sein
wird, von ihm als 'ontologisch' bezeichnete Entgegensetzung von
Individuum und Gesellschaft in seiner Theorie eingebaut, nicht nur auf
der Ebene einer Sozialisation der Individuen, die er ja als familiäre
Interaktion aufzeigen müßte, dessen er sich, wie gesehen, oberflächlich
genug erledigt; sondern auch auf dem Hintergrund von 'Seinsebenen',
die nur durch den gemeinsamen Charakter des magmatischen
Imaginären miteinander verbunden sind, denn "das radikale Imaginäre
existiert als Gesellschaftlich-Geschichtliches und als Psyche-Soma. Als
Gesellschaftlich-Geschichtliches ist es offenes Strömen des anonymen
Kollektivs; als Psyche-Soma ist es Strom von Vorstellungen/Affek-
ten/Strebungen"[93].

Castoriadis begreift aber die Seinsweise dieses Imaginären, zum
Beispiel die "Seinsweise des Unbewußten"[94], nicht etwa im Sinn eines
aktuellen Existenzmodus einer bestimmten Form, sondern als ontologi-

sche Bestimmtheit; das Imaginäre nimmt einen ontologischen Status an und ist identisch mit *elan vital* und 'Leben' denn Castoriadis ist davon überzeugt, "daß nämlich die Imagination radikal bildend ist, nicht *Einbildungskraft* (i. O. deutsch), sondern *Bildungskraft* (i. O. deutsch); daß sie Bilder und Formen schafft (...) Wahrnehmung gibt es nur, weil es den Vorstellungsstrom gibt. So gesehen ist das Imaginäre — als gesellschaftliches Imaginäres und Imagination der Psyche — ebenfalls logische und ontologische Bedingung des 'Realen'"[95]. Die logische Bedingung des Imaginären für die Existenz und Form des Realen ist das schon angesprochene Symbolische, das Castoriadis durch das 'Imaginäre' ersetzt und das als Konstituierung von Welt dem Realen 'vorausgehen' muß, ihm vorgängig ist, den logischen und damit für Castoriadis ebenfalls ontologischen Primat besitzt. Was Castoriadis mit der ontologischen Bedingung des Realen im Imaginären meint, muß aus der dem Imaginären zugeschriebenen Aktivität, Spontaneität, Unbedingtheit begriffen werden: Castoriadis benötigt zur Konzeption seiner Gesellschaftstheorie als einer des instituierten Imaginären, der imaginären Institution, mit der Bedeutung, die er dem Imaginären zuweist, notwendig ein Imaginäres, das eine *prima causa* darstellen kann und damit eine Art Seinsgrund, der allem zugrundeliegt: ein ontologisches *perpetuum mobile*. Diese Argumentationsstruktur ist merkwürdig vertraut aus der philosphischen Tradition, denn Castoriadis muß dabei genauso vorgehen wie Descartes: wenn er an einen Punkt gelangt, an dem er — nachdem er wieder einmal 'nachgewiesen' hat, daß die traditionelle und auch die moderne strukturalistische Denkweise nichts erklären — eine neue Erklärung oder Konzeption entwickeln müßte, zieht er wie Descartes Gott aus dem Ärmel, nämlich eben das radikale Imaginäre, das zur Erklärung für alles bisher Ungelöste herhalten muß und alle Zweifel im Vertrauen in seine Kraft wie bei Descartes in die Gutmütigkeit Gottes aufhebt; Castoriadis' abenteuerliche Entdeckungsfahrt besteht in der jubelnden Begrüßung eines jeden neuen Gestades, das sich bald darauf als die schon bekannte Küste herausstellt. Das Imaginäre erfüllt in seiner theoretischen Organisation die Funktion Gottes im traditionellen philosphischen und ideologischen Diskurs: der irgendwo im Sein fest verankerte Haken, an dem der jeweilige theoretische Anzug aufgehängt werden kann; der Platzhalter Gottes für Castoriadis ist das radikale Imaginäre, von dem ebenfalls keiner genaueres weiß, denn "diese Formen (der gesellschaftlichen poiesis — FL) lassen sich nicht einfach nach den bisherigen Kriterien der instituierten Vernunft diskutieren und bewerten"[96] — ein gefährlicher Schlenker ins Halbmythische, der offen läßt ob es sich um eine Kritik der instrumentellen Rationalität als Herrschaftsinstrument oder als Entbindung daseinshafter Uneigentlichkeit handelt, die nach bewegter Eigentlichkeit schreit und auch eine Suspension der Vernunft bedeuten kann, wie sie historisch schon erlebt wurde und ein erneuter Beleg für Castoriadis' mögliche Integrierbarkeit ins konservative post-histoire ist.

Als "die Wirkung, die ihre Ursache überschreitet"[97] wird das Imaginäre hier jedenfalls als Transzendentes zum Transzendentalen, eine *prima causa* als *causa sui* des Gesellschaftlichen; die Gesellschaft wird in Analogie zum bürgerlichen Geniekult zum Künstler, zur permanenten Schöpfung in der Geschichte durch das Imaginäre, denn "die Geschichte ist wesentlich *poiesis*, und zwar nicht nachahmende Poesie, sondern ontologische Schöpfung und Genese im und durch das Tun und das Vorstellen/Sagen der Menschen"[98]. Gegen Castoriadis' eigentliche Intention erwecken solche Formulierungen Anklänge an die hymnische Euphorik praktizierter Lebensphilosophie deutscher Provenienz auf dem Monte Veritá und deren quasisoziologische, politische Verwandlung in einen magischen Futurismus á la Jünger[99]; der Grund für diese fatale und ungewollte Nähe der Formulierung zum präfaschistischen Weltanschauungsbrei[100] in der unseligen Konstruktion des Imaginären, das als gesellschaftlicher *élan vital* hilflos seiner theoretischen Vermarktung für alles und jedes zusehen muß, so wie Bergsons kosmologisches Magma, politisch uminterpretiert, dem faschistischen Vulkanausbruch zutrieb.

In seiner Bemühung, das Subjekt mit den falschen Mitteln zu retten, muß Castoriadis so weit gehen, bis er zur Ontologie gelangt, in der der Subjektivität in Form des Imaginären die Seinsstufe reserviert ist, die allen anderen Stufen zugrundeliegt: "die Vorstellung ist fortwährendes Anwesendseinlassen, unaufhörliches Fließen, in und mit dem alles ist, was es auch sei. Sie gehört nicht zum Subjekt, sie *ist* das Subjekt (...) jenes dicht und kontinuierliche Strömen..., das wir sind (...) immanentes Transzendieren, Über-sich-hinaus-gehen zum anderen"[101]. "Die Souveränität der radikalen Imagination"[102] ist so unantastbar, so absolut, daß es der Urgrund alles Seienden ist, denn zwar muß

das Imaginäre...das Symbolische benutzen, nicht nur um sich 'auszudrücken' – das versteht sich von selbst – sondern um überhaupt zu 'existieren', um etwas zu werden, das nicht mehr bloß virtuell ist (...) In dem Maße jedoch, wie das Imaginäre letztlich auf eine ursprüngliche Fähigkeit zurückgeht, sich mit Hilfe der Verstellung ein Ding oder eine Beziehung zu vergegenwärtigen, die nicht gegenwärtig sind (die in der Wahrnehmung nicht gegeben sind oder es niemals waren), werden wir von einem letzten oder *radikalen Imaginären* als der gemeinsamen Wurzel des *aktualen* Imaginären und des *Symbolischen* sprechen. Es handelt sich dabei um die elementare und nicht weiter zurückführbare Fähigkeit, ein Bild hervorzurufen[103].

Hier wandert Castoriadis' Theorie ab in eine lebensphilosophische Ontologie, die den wertvollen Begriff des Imaginären so ausdehnt, daß er stumpf wird; Castoriadis muß zurückgehen auf das, was er dauernd angreift: das absolute, souveräne fichtesche Subjekt, wie sich aus einer Fußnote zum obigen Zitat ergibt, wo er den jungen Hegel der Jenenser Realphilosophie von 1805/06 zitiert:

der Mensch ist die leere Nacht, dies leere Nichts, das alles in ihrer Einfachheit enthält, ein Reichtum unendlich vieler Vorstellungen, Bilder, deren keines ihm gerade einfällt oder die nicht als gegenwärtige sind. Dies (ist) die Nacht, das Innre der Natur, das hier existiert – *reines Selbst*. In phantasmagorischen Vorstellungen ist es ringsum Nacht: hier schießt denn ein blutig Kopf, dort ein(e) andere weiße Gestallt plötzlich hervor und verschwindet ebenso. Diese Nacht erblickt man, wenn man dem Menschen ins Auge blickt – in eine Nacht hinein, die *furchtbar* wird; es hängt die Nacht der Welt hier einem entgegen. – *Macht, aus dieser Nacht die Bilder hervorzuziehen oder sie hinunterfallen zu lassen: Selbstsetzen, innerliches* Bewußtsein, *Tun, Entzweien*[104].

Castoriadis verschmilzt das Imaginäre als gesellschaftlichen Bereich mit dem Akt der Setzung selbst, also mit der Negation des unmittelbar Positiven, der Schöpfung: imaginär ist alles, was nicht scheinbar unmittelbar ist, was nicht 'Natur' ist; die Gesellschaft als solche ist das Imaginäre, und der Begriff verliert seine kritische Schärfe.

Von dieser Überlegung aus wird die entscheidende Macht des Imaginären über das Symbolische verständlich; der Symbolismus setzt die Fähigkeit voraus, zwischen zwei Termini ein dauerhaftes Band herzustellen, so daß der eine den anderen 'vertritt'. Doch erst auf sehr weit fortgeschrittenen Stufen des klaren rationalen Denkens werden Einheit und Unterschiedenheit dieser drei Elemente (Signifikant, Signifikat und ihre Verbindung *sui generis*) in ihrer festen und zugleich geschmeidigen Beziehung erkannt. Geschieht das nicht, so regrediert die symbolische Beziehung – deren 'eigentliche' Verwendung die imaginäre Funktion *und* deren Beherrschung durch die rationale Funktion voraussetzt – oder bleibt überhaupt, was sie ursprünglich war: ein starres Band zwischen Signifikant und Signifikat, Symbol und Ding; eine Verbindung, die meist als Gleichsetzung, Teilhabe oder Verursachung gedacht wird. Mit anderen Worten, man bleibt im aktualen Imaginären[105].

Das aktuale Imaginäre bezeichnet also in etwa die Verwendung des
Begriffs des Imaginären, den ich oben vorgeschlagen habe: die
Verdinglichung des Symbolischen und damit auch die des Bewußtseins;
um aber der Ontologie Imaginären zu entgehen, muß man die hier
aufgezeigte Beziehung von Imaginärem und Symbolischem anders
konzipieren als Castoriadis, damit dem Imaginären nicht das ontologische Prius aufgebürdet wird.

Der Symbolismus, das Symbolische bedarf nicht des radikalen
Imaginären, um zur Existenz zu kommen: er ist selbst, setzt nicht
voraus, die Fähigkeit, zwischen zwei Termini ein dauerhaftes Band
herzustellen. Dieses Band wird erst dann fix und rigide, das so als
Kombination von Signifikant und Signifikat entstandene Zeichen erst
dann pathologisch, wenn diese Beziehung als willkürliche, arbiträre
nicht bewußt ist, das heißt wenn der Symbolismus als System von
Zeichen den Charakter des Naturwüchsigen annimmt, ideologisch wird,
sich verselbständigt, wenn man so will; in dieser Struktur hat sich das
Imaginäre des Symbolischen bemächtigt und es zu einem Imaginären
gemacht. Das kann nur deswegen überhaupt geschehen, weil das
Symbolische zwei konstitutive Momente besitzt: das 'emphatische', das
seine reale — das heißt kommunikative — Bedeutung betrifft: ein
utopisches Potential von Kommunikation in der Struktur von Sprache
selbst, und funktionale, das die semiotische Funktion der 'Ordnung des
Diskurses' bezeichnet, die reine Funktion des Symbolischen als
Zeichensystem zur Übermittlung von Informationen ohne normativen
Gehalt — eine Einbahnstraße, bestenfalls Wechselwirkung der Informa-
tionsübermittlung durch ein konstituiertes Zeichensystem auf der
einen, eine interaktive Dialektik der Anerkennung in der Kommunikation auf der anderen Seite. Das funktionale Moment wird, wenn es vom
'emphatischen' isoliert bleibt, vom Imaginären 'erobert' indem dieses
die Trennung instituiert, genauer: selber ist; das Imaginäre ist die
Abspaltung des utopischen Charakters am Symbolischen; da das
Symbolische reale Wirkung hat — indem es 'Welt' konstruiert, den
Diskurs des Sozialen ordnet, das Subjekt konstituiert — gewinnt diese
auch das Imaginäre als Verdinglichung des Symbolischen.[106]

Castoriadis fällt durch die mangelnde Differenzierung innerhalb der
zentralen Begriffe seiner Theorie der Unfähigkeit der Erklärung
gesellschaftlicher Verdinglichung anheim; weil er sich in einem
Teufelskreis bewegt, da ja seine Voraussetzung zur Verhinderung oder
Überwindung des 'aktualen Imaginären', also zur Realisierung des

Symbolischen im Sinn der rationalen Reflexion auf dessen Bedingungen und Struktur 'auf sehr weit fortgeschrittenen Stufen des klaren rationalen Denkens' ohne das Symbolische nicht zu leisten ist; um auf diese 'Stufen' zu gelangen, müßte sich ja immerfort das Imaginäre wie von selbst verflüchtigt haben, damit das Symbolische nicht in ihm hängenbliebe.

Wird umgekehrt das Symbolische als die Dimension der Konstitution von Welt durch Worte, Bilder und Strukturen und das Imaginäre als die Verdinglichung dieser Dimension, das heißt als die bloß formale Organisation dieses prinzipiell, von der Anlage her kommunikativen Potentials betrachtet: eine rein mechanische, automatische Reduktion von Komplexität nach selektivem Muster, so wird alles erheblich plausibler; es ließe sich zudem beziehen auf und anschließen an den symbolischen Interaktionismus und damit auf die Dimension, die bei Castoriadis wesentlich fehlt: Intersubjektivität.

> Natürlich sind die Bedürfnisse im gesellschaftlich-geschichtlichen Sinne – also nicht im Sinne biologischer Erfordernisse – ein Produkt des radikalen Imaginären. Das 'Imaginäre', das die ausbleibende Befriedigung dieser Bedürfnisse kompensiert, ist also nur ein sekundäres und abgeleitetes Imaginäres. Das trifft auch bestimmte psychoanalytische Strömungen der Gegenwart, denen zufolge das Imaginäre ein ursprüngliches Aufklaffen, eine ursprüngliche Spaltung des Subjekts 'vernäht'. Aber dieses Aufklaffen verdankt sich selbst nur dem radikalen Imaginären des Subjekts[107].

Hier kann man sehen, wie Castoriadis — gegen seine Absage an eine grundsätzliche Entfremdung durch Sprache oder Gesellschaft[108] — dennoch gezwungen ist, dieser Ontologie Lacans zu folgen: hätte er die Produktion von Bedürfnissen als Ergebnis des Symbolischen betrachtet und genauer die Produktion von Bedürfnissen, die nicht befriedigt werden können, sondern imaginär kompensiert werden müssen, als Ergebnis eines durchs Imaginäre verdinglichten Symbolischen, wie es in einer Gesellschaft mit Herrschaft vorfindbar ist, so könnte er der Ontologie dieses Aufklaffens entkommen durch die Analyse dieses imaginären Symbolischen, statt die Fähigkeit zum radikalen Imaginären vor aller symbolischen 'Prägung' ansetzen zu müssen, also ein anthropologisch-ontologisches Fixum vor jeglicher Sozialisation anzunehmen'.

Da der Eindruck entsteht, daß Castoriadis durch die Pose des Spontanen, das als vorgängig zum Symbolischen posiert, hinter Hegel zurückfällt — dessen Leistung ja gerade die gesellschaftliche Rekonstruktion dieses

Spontanen und Subjektiven war — fragt sich, was eigentlich inhaltlich neu ist bei Castoriadis, wenn er das Imaginäre — das eigentlich das Symbolische ist — als den Quell und die Ordnung aller unserer 'Weltanschauung' zeigt. Die Naivität, mit der Castoriadis das ganze Buch hindurch vorgeht, als ob es eine philosophische Tradition nach Aristoteles einfach nicht mehr gäbe, wird allerdings verstehbar, wenn er am Ende eingeht auf die wohl der ganzen Anstrengung zugrundeliegende "Frage von ungeheurer Tragweite... Muß man sich mit der bloßen Feststellung der Grenzen dieser (Identitäts — FL) Logik und der ihr wesensgemäßen Ontologie begnügen — oder kann man diese einfache negative Ontologie überwinden und einen Weg (oder mehrere) freimachen, der es gestattet, das Seiende zu denken, ohne sich dabei auf Angaben darüber beschränken zu müssen, wie es nicht zu denken sei"[109]. Das ist der Punkt, der ihn beschäftigt: wie zu denken sei — der negativen Dialektik, die sich mit der dialektischen Kritik der Identitätslogik in der Immanenz herumzuschlagen hat, eine positve Bestimmung des Seienden ohne traditionelle Ontologie entgegenzustellen, möglichst ohne sich dabei die Hände schmutzig zu machen an Paradoxien. Bei dieser Fragestellung kann er jedoch nur — und gerade — ins von der Ontologie aufgestellte Messer laufen, selbst wenn er noch so oft 'gesellschaftlich-geschichtlich' schreibt: in dem Moment, in dem Gesellschaft als Seiendes gedacht wird, ist ja die Falle schon aufgestellt, weil ein 'Sein' sprungbereit im Hintergrund lauert; ihm selbst bleibt gar nichts anderes übrig, als sich auf den Weg zu machen und das Imaginäre — als Vermögen, ganz allgemein, als zugleich Positives: Bestimmbares und Posierendes, Positivierendes: Schöpfendes, Setzendes — als den Ausweg aus dem Dunkel zu sehen, der wie in Jules Vernes Reise zum Mittelpunkt der Erde nur der auf den Fluten des Vulkans sein kann: selbsttätiges Magma, ein Gott der Geschichte, der alle Bestimmungen des Gottes des Seins der Tradition besitzt.

Castoriadis muß, genau wie Descartes, um zum Positiven zurück-kehren zu können, einen Punkt absoluter Versicherung finden und als dessen Absicherung einen Gottesbeweis führen, der wie alle Gottesbe-weise bei der Postulierung hängenbleibt: "Wir behaupten, daß alles potentiell Gegebene — Vorstellung, Natur, Bedeutung — von der Seinsart des *Magma* ist"[110]. Er muß, weil er von Anfang an die falsche Konzeption verfolgt hat, die Geburt eines neuen Mythos, einer neuen Religion inszenieren, die er schwärmerisch feiert: das Imaginäre, die Vorstellung "ist das, was uns im Licht stehen läßt, selbst wenn wir die

Augen schließen, wodurch wir Licht in der Dunkelheit sind und wodurch *selbst der Traum noch Licht ist*"[111] — ein Licht, das sich eher der Metapher der Offenbarung oder gar der der Erleuchtung als der der Aufklärung verpflichtet fühlt.

Theologie des Imaginären

Und dennoch scheint der Vorwurf einer Gesellschaftsontologie, einer Ontologisierung der Gesellschaft Castoriadis nicht völlig gerecht zu werden, zumal er es ablehnt, wie Heidegger "aus der ontologischen Differenz einen absoluten Unterschied" zu machen, also das Sein vom Seienden getrennt zu denken[112]; somit scheint er gegen losgelöste, unhistorische, ontologische Fragestellungen gefeit. Aber ironischerweise läuft er gerade durch das Insistieren auf dem Zusammenhang von Sein und 'konkretem' Seienden der Ontologie ins Messer, denn er glaubt — oder besser: wird zu glauben genötigt —, einen Seinsgrund zum Begreifen dieses Seienden zu benötigen, den er natürlich im Imaginären findet, das so zur ontologischen Struktur und Konstante gerät, die die Änderungen und das heißt die jeweiligen Formen des Seienden bestimmt.

Castoriadis' Theorie gleitet deshalb den Abhang zum Mythos hinunter — genauer: wird ihn hinabgezogen —, weil sie mit 'mythischen' 'Kategorien' arbeiten muß: "Frage nach der Geschichte, Frage nach der Wahrheit, Frage nach dem Verhältnis beider. Eine philosophische (und im strengen Sinn des Worts politische) Frage, die ausgegrenzt wird, wenn man die Wissenschaft zur bloßen Aufeinanderfolge von 'Paradigmen' degradiert oder sich auf die Beschreibung dessen beschränkt, was man die 'Episteme' einer jeden Epoche genannt hat..."[113]. Geschichte und Wahrheit werden so zu Entitäten, zu mystischen Kategorien, jenseits aller — angeblich verdinglichenden — Klassifikation, denn "die Geschichte, auch die Geschichte des Denkens, ist im vollem Sinne eine ontologische Schöpfung. Sie ist nicht bloß Produktion (also Reproduktion von Exemplaren eines gegebenen *eidos*) noch bloße ontologische Erschaffung, Hervortreten eines anderen *eidos*. Was sie erschafft, sind vielmehr *eidos* Typen, eine andere Definition von Figur und Hintergrund, eine andere Zugehörigkeit/Verschiedenheit ihrer 'Bestandteile' "[114]. Es bleibt eigentlich unklar, wie Castoriadis das Subjekt gegen die strukturalistische Theorie retten will, wenn 'die

Geschichte' die einzige ist, die Aktivität, Schöpfung zeigen kann, da ja alles andere bloß technische Reproduktion oder Ausbau schon von der Geschichte hergestellter, imaginierter *eidos*-Typen ist. Natürlich würde Castoriadis nicht leugnen, daß die Geschichte von Menschen gemacht wird — das wieder gegen Althusser ins Bewußtsein bringen zu müssen ist ja seine Absicht, der Impuls seiner Theorie. Aber durch seine Konzeption dieser Geschichte des Imaginären ist er permanent gezwungen, die 'Gesellschaft, die sich imaginär instituiert', wie eine Person, ein Ich zu sehen, das sich eine 'Form' gibt — ein der idealistischen Philosophie analoger Bildungsprozeß[115]. Statt der marxschen Trennung von menschlicher Geschichte und Naturgeschichte zu folgen, derzufolge wir die eine gemacht haben und die andere nicht, führt Castoriadis beides zusammen durch die Kraft des Imaginären als Motor der Geschichte, die so als Naturgeschichte erscheint, als eine mit deren Schöpfung die Menschen wenig anderes zu tun haben als die Reproduktion der Formen, die den jeweiligen Gesellschaften zugrundeliegen. Gesellschaft wird eine Natur mit menschlichen Eigenschaften: diese Anthropologisierung der Gesellschaft und Personalisierung der Geschichte ist nicht nur mythisch und — wie gesehen — politisch bedenklich, sondern versagt auch einfach dann, wenn es eben diese *eidos*-Typen also die ureigensten Produkte des 'Imaginären' zu erklären gilt.

Castoriadis entwickelt die 'Ordnung des Diskurses'[116] also die Organisation des sozialen Raums, gesellschaftlichen Felds, des Symbolischen als "*Legein*: unterscheiden/auswählen/aufstellen/zusammenstellen/zählen/sagen; zugleich Bedingung und Schöpfung der Gesellschaft, von Bedingtem geschaffene Bedingung. Damit die Gesellschaft existieren kann, damit eine Sprache eingeführt werden und funktionieren kann, damit sich schließlich eine reflektierte Praxis entfalten kann und die Menschen nicht mehr gezwungen sind, sich bloß in der Phantasie aufeinander zu beziehen, muß in einer bestimmten Weise auf einer bestimmten Ebene oder Schicht des gesellschaftlichen Handelns und Vorstellens alles so lange zurechtgestutzt werden, bis die Cantorsche Definition paßt"[117]; gemeint ist die Definition der Menge als das vereinigte und vereinigende Ganze ihrer Elemente, geschaffen durch die Produktion eines "Schemas der Trennung", dem natürlich ein "Schema der Vereinigung"[118] entsprechen muß. "Identität und Differenz"[119] werden instituiert, wie Castoriadis sagen würde, oder konstituiert und als Diskursregel konstruiert. Aber Castoriadis muß diese gesellschaftliche Handlung so denken, daß sie "mit einem Schlag"[120] geschieht; aus der gesellschaftlichen Schöpfung wird eine fichtesche Setzung, bei der das jetzt Gesellschaft gewordene Ich — das aber alle Ichqualitäten beibehält — mit einem Schlag das Nicht-Ich setzt und zugleich so organisiert, daß es als Moment der Gesellschaft erscheinen kann, um damit die klassische idealistische These zu realisieren, daß Identität die Identität von Identität und — gesellschaftlich organisierter — Nicht-Identität ist. Dieser Setzungsakt — von dem man nicht weiß, wie er in einer Gesellschaft geschieht, geschehen soll und wer ihn vollzieht: — ist das radikale Imaginäre, das bei Castoriadis Gefahr läuft, selber ungesellschaftlich, ungeschichtlich zu werden, da dessen Konstitutionsbedingungen nicht ersichtlich sind. Die Geschichte wird Subjekt, die Gesellschaft Individuum, Person nach dem Modell der Psyche, angetrieben und motiviert von Gesellschaftstrieben: sie instituiert ihr Imaginäres, das sie erst als Gesellschaft auftreten läßt, wie das Ich das Nicht-Ich setzt, demgegenüber es sich sich seiner selbst erst bewußt werden kann als absolute Macht. Castoriadis ist auf der Suche nach dem Absoluten, das im System seiner Philosophie noch seinen festen, systematisch notwendigen; zwar gesellschaftlich uminterpretiert en aber unentbehrlichen Platz hat: das Imaginäre als Setzung ist so nah bei Fichte, daß er "absolute Schöpfung"[121] dazu sagen muß.

Aber Castoriadis geht letztlich so weit, diese Gleichsetzung der Gesellschaft mit dem bürgerlichen Phantasma vom Ich noch zu überbieten durch den Rückgriff auf das, von dem die bürgerliche Philosophie sich zu emanzipieren bemüht war; seine Gesellschaftstheorie wird Theologie. Das legein,

der Symbolismus ist wohl oder übel auf etwas Außersymbolisches bezogen, was aber auch nicht bloß Reales oder Rationales ist. Dieses Element, das der Funktionalität jedes institutionellen Systems eine besondere Ausrichtung gibt und das die Wahl und die Verknüpfungen der der symbolischen Netze überdeterminiert; dieses Element, das sich jede geschichtliche Epoche schafft und in dem sie unnachahmlich ausdrückt, wie sie ihre eigene Existenz, ihre Welt und ihre Beziehungen zu dieser erlebt, dieses zentrale Bedeutete/Bedeutende, diese Quelle allen unzweifelhaften und unbezweifelbaren Sinns, dieser Träger aller Verknüpfungen und Unterscheidungen zwischen Wichtigem und Unwichtigem, dieser Ursprung der Seinsbereicherung, den die Gegenstände praktischer, affektiver oder intellektueller, individueller oder kollektiver Besetzungen erfahren — dieses Element ist nichts anderes als das *Imaginäre* der Gesellschaft oder der jeweiligen Epoche"[122]

— und hat somit nahezu alle Bestimmungen von Gott: Ursprung, Quelle, Bedeutetes/Bedeutendes, Richtungsgeber, Sinnschöpfer, das, von dem alles ausgeht und auf das alles hinläuft: Alpha und Omega. Es ist nicht einsichtig, warum sich das Symbolische auf etwas Demiurgisches außerhalb seiner selbst beziehen muß, statt auf eine gesellschaftliche Realität, zu der es eine dialektische Beziehung unterhält, funktional von ihr abhängt und sie dennoch 'formt'; Castoriadis aber ordnet dieser noch das Imaginäre vor, das letztlich ebenso unzugänglich sei wie Gott, denn

erfaßt werden können sie (die gesellschaftlich imaginären Bedeutungen -FL) nur in abgeleiteter und mittelbarer Weise, nämlich als evidenter und doch nie genau zu bestimmender Abstand zwischen dem Leben und der tatsächlichen Organisation einer Gesellschaft einerseits und der ebenso undefinierbaren, streng funktional-rationalen Organisation dieses Lebens andererseits. Danach wären die gesellschaftlichen imaginären Bedeutungen als 'kohärente Deformation' des Systems der Subjekte, Objekte und ihrer Beziehungen zu verstehen, als die jedem gesellschaftlichen Raum eigentümliche Krümmung, als der unsichtbare Zement, der den ungeheuren Plunder des Realen, Rationalen und Symbolischen zusammenhält, aus dem sich jede Gesellschaft zusammensetzt – und als Prinzip, das die dazu passenden Stücke und Brocken auswählt und angibt. Die gesellschaftlichen imaginären Bedeutungen – sofern es wirklich die letzten sind – *denotieren* nichts, *konnotieren* aber fast alles"[123].

Gott wird sichtbar nur in seinen Werken: das Imaginäre wird ein Mystisches, das wie der Rauch unter Pythias Dreifuß aus den uneinsichtigen und tödlichen gesellschaftlichen Abgründen und Spalten aufsteigt, unter denen heißes Magma fließt, das sich orakelhaft in Bedeutungen umsetzt und niederschlägt, kondensiert als nebelhafter Dampf.

Hier zeigt sich das zentrale methodische Problem von Castoriadis: da ihm ein Begriff von gesellschaftlicher Totaltität, der das Konzept der Intersubjektivität notwendig einschließen würde, fehlt, muß er den Konflikt von System und Lebenswelt, der sich zu Beginn des obigen Zitats ausdrückt, mythisch fassen als den zwischen radikalem Imaginären und aktualen Imaginären, das heißt dem verdinglichten Imaginären, das seine Bewegungsfreiheit verloren hat. Anstatt die Verdinglichung als die Funktionalisierung des lebensweltlichen Symbolischen durch ein formal organisiertes Imaginäres zu interpretieren, also die schrittweise Eroberung der Lebenswelt, die sich auf eine in der Lebenswelt vorhandene Struktur des Imaginären stützen kann, zu sehen, läßt Castoriadis dies als Kampf im Imaginären — in meinem Sinn: im Symbolischen — allein erscheinen. Das ungeheuer Fruchtbare dieses Ansatzes: nämlich die Möglichkeit des Imaginären zu zeigen, das Symbolische zu erobern, indem dessen kommunikative Dimension abgespalten wird und unrealisiert bleibt, verschenkt Castoriadis durch seine Definition und Konzeption des Imaginären als Kreation, Schöpfung. Dadurch ergeben sich weitere methodische Mängel, die Gesellschaft zu denken, die ihn zwangsläufig der Ontologie in die Arme treiben, da er keinen Begriff von Intersubjektivität besitzt, der es ihm ermöglichte, der Konstruktion des Imaginären als demiurgischem Gesellschaftstrieb zu entgehen. Konsequenterweise findet so die Dimension symbolisch konstituierter Realität so gut wie keine Beachtung, obwohl gerade hier die entscheidende gesellschaftstheoretische Analyse zu leisten wäre, in der die einzelnen gesellschaftlichen Strukturen aufzuzeigen wären. Castoriadis ist stattdessen interessiert am imaginären Magma aller Gesellschaften und der Geschichte, und vergibt so die Möglichkeit, Genaueres über das Imaginäre in der Moderne zu sagen. Da ein Begriff wie der der Moderne in Castoriadis' Theorie keinen systematischen Stellenwert besitzt und er somit die gerade für die Moderne konstitutive Rolle des Imaginären nicht in den Blick — geschweige denn auf den Begriff — bringen kann, gerät sein Imaginäres zur überhistorischen, ontologischen Figur einer Theorie des Imaginären, die sich als gesellschaftsphilosophische Rede vom gesellschaftlichen Sein und Seienden mißversteht.

Die Theorie des Imaginären lernt somit von Castoriadis im wesentlichen auf negativem Weg: das Imaginäre kann, selbst wenn es konstitutiv für die moderne Gesellschaft ist, nicht den Platz des traditionellen Subjekts einnehmen und Motor der Geschichte sein. Es muß vielmehr, statt als *Unbedingtes* in der Theorie aufzutreten, durch seine *Funktion* charakterisiert werden, die es in ein funktionales, das heißt strukturelles Verhältnis zur Gesellschaftsformation bringt, aus dem heraus es begriffen werden muß; dies bedingt einen methodischen Ansatz, der den Funktionsbegriff weder ganz noch tendenziell — wie Castoriadis — aufgibt. Anstatt das Imaginäre als *Akt*, schon gar der Schöpfung, zu verstehen, muß es vielmehr als *Struktur* begriffen werden, die wirkt und Effekte erzielt, nicht handelt; damit geht einher, daß eine Theorie der Gesellschaft ohne Ebenen der Vermittlung nicht konstruierbar ist; die Intersubjektivität ist ein unverzichtbarer systematischer Bestandteil einer Theorie des Imaginären, die den Standpunkt der Handlungs- und der Systemtheorie zugleich einzunehmen hat, um aufzuzeigen, daß das Imaginäre als formal organisierte Sphäre die Verzerrung des Symbolischen in der Form der Abspaltung des in seine Struktur eingebauten kommunikativen Potentials ist.[124]

Nagelschmiedshütte 10
5000 Köln 40

ANMERKUNGEN

Nach Fertigstellung des Manuskripts 1985 sind zum Thema noch erschienen: Jürgen Habermas, *Exkurs zu Castoriadis*: 'Die imaginäre Institution', in: *Der philosophische Diskurs der Moderne*, Zwölf Vorlesungen, Ffm. 1985, 380-390; und Axel Honneth, *Eine ontologische Rettung der Revolution. Zur Gesellschaftstheorie von Cornelius Castoriadis*, in: *Merkur* 9/10, 1985, 807-822.

[1] Castoriadis, [1984; 238].
[2] *Ibid.*, 95.
[3] *Ibid.*, 40.
[4] *Ibid.*, 373.
[5] *Ibid.*, 179 Fußnote.
[6] *Ibid.*, 185.
[7] Castoriadis, [1981; 13].
[8] *Ibid.*, 36.
[9] *Ibid.*, 34.
[10] *Ibid.*, 77.
[11] *Ibid.*, 31.
[12] *Ibid.*, 37.
[13] *Ibid.*, 38f.
[14] *Ibid.*, 40ff.
[15] *Ibid.*, vgl. 194ff.
[16] *Ibid.*, 608.
[17] *Ibid.*, 362.
[18] *Ibid.*, 196.
[19] *Ibid.*, 175.
[20] *Ibid.*, 216.
[21] *Ibid.*, 217.
[22] Vor allem: Morris [1973; 288-323, bes. 301-318].
[23] vgl. Rossi-Landi [1981; 235-267].
[24] U.a.: Baudrillard [1982].

168

24 Castoriadis [1981, 15].

26 *Ibid.*, 417.

27 *Ibid.*, 199.

28 *Ibid.*, 200.

29 *Ibid.*, 214.

30 *Ibid.*, ebd.

31 *Ibid.*, 604.

32 Foucault [1977].

33 Castoriadis [1981; 386].

34 *Ibid.*, 385.

35 *Ibid.*, 395.

36 Es würde zu weit führen, hierzu ausführliche Literaturangaben zu geben; dieses Thema hat sich von Kant über Wittgenstein in viele Aspekte zersplittert.

37 *Labyrinth* 92; mit diesen Anspielungen ist Lacan gemeint und es ist hier vielleicht eine kurze Fußnote zum Verhältnis Castoriadis' zu Lacan vonnöten. Castoriadis übernimmt einfach von Lacan, ohne das nachzuweisen oder bloß darauf hinzuweisen, ja sogar noch mit der Attitude eines Arguments gegen Lacan [vgl. 1981; 97f]; er 'folgt' Lacan in der Interpretation des Ödipuskomplexes – sowie in der ganzen Sozialisationstheorie (490 bis 530) –, wie er selber zugibt (511f. und Fußnote). Castoriadis' 'direkte' Auseinandersetzung mit Lacan – im Aufsatz in 'Durchs Labyrinth' – erfolgt anhand eines Buches über die Lacan-Szene mit wenig Argumenten gegen Lacans Theorie.

38 *Ibid.*, 192.

39 *Ibid.*, 532 Fußnote.

40 *Ibid.*, 513.

41 *Ibid.*, 174f.

42 *Ibid.*, 178f.

43 *Ibid.*, 409.

44 *Ibid.*, 399 Fußnote.

45 *Ibid.*, 399.

46 *Ibid.*, 375 und 377.

47 *Ibid.*, 77.

48 *Ibid.*, 12.

49 *Ibid.*, 83.

50 *Ibid.*, 78.

51 *Ibid.*, 81.

52 *Ibid.*, Ebd.

53 *Ibid.*, 84.

54 *Ibid.*, 413; man achte auf den heideggerschen Sinn dieses Sich-seinlassens; dazu auch 287 und 436.

55 *Ibid.*, 245.

56 *Ibid.*, 221.

57 *Ibid.*, 203.

58 *Ibid.*, 210f. Fußnote.

59 *Ibid.*, 595.

[60] *Ibid.*, 591f.
[61] *Ibid.*, 598.
[62] *Ibid.*, 602.
[63] *Ibid.*, 382.
[64] *Ibid.*, 383.
[65] Vgl. dazu Honneth [1977; 433, 437f].
[66] Castoriadis [1981; 12].
[67] *Ibid.*, 286, Fußnote 17.
[68] *Ibid.*, 165.
[69] *Ibid.*, 165f.
[70] *Ibid.*, Vgl. 166f.
[71] *Ibid.*, 169.
[72] *Ibid.*, 256 Fußnote
[73] *Ibid.*, 397.
[74] Habermas [1974; 44f, 1976; 103f].
[75] Vgl. Hegel [1971; 101].
[76] Vgl. Fußnote 22. Zür Ästhetisierung der Realität vgl. Löwenich [1987; 124-130, 1986b; 143-163, 1988a; 85-91].
[77] Castoriadis [1981; 273].
[78] *Ibid.*, 414.
[79] Vgl. dazu Luhmann [1964, 1969].
[80] Castoriadis [1981; 42].
[81] *Ibid.*, 457.
[82] *Ibid.*, 459f.
[83] *Ibid.*, 456.
[84] *Ibid.*, 10.
[85] *Ibid.*, 470f.
[86] *Ibid.*, 471.
[87] Hegel [1971; 21].
[88] Castoriadis [1981; 476].
[89] *Ibid.*, 482f.
[90] *Ibid.*, 456.
[91] *Ibid.*, 459.
[92] *Ibid.*, 457.
[93] *Ibid.*, 493.
[94] *Ibid.*, 603.
[95] *Ibid.*, 457.
[96] *Ibid.*, 553f.
[97] *Ibid.*, 15.
[98] *Ibid.*, 39.
[99] *Ibid.*, 13.
[100] Vgl. dazu Pfankuch, Lepper [1981].
[101] Vgl. dazu Glaser [1979].
[102] Castoriadis [1981; 546].

170

103 *Ibid.*, 505.

104 *Ibid.*, 218.

105 Zit. bei Castoriadis, [1984; 218f]. Fußnote. Zun Hintergrund dieses Zitats vgl. F. Löwenich [1986; 53-57].

106 Castoriadis [1984; 218f].

107 Vgl. da zu den 'Gutfurfeiner Theorie des Imaginären' in: Löwenich [1990; 211-267].

108 Castoriadis [1984; 228f]. Fußnote.

109 *Ibid.*, vgl. 194f.

110 *Ibid.*, 559.

111 *Ibid.*, 565.

112 *Ibid.*, 546.

113 Castoriadis [1981; 132, 153f].

114 Castoriadis [1981; 12].

115 *Ibid.*, 20f.

116 *Ibid.*, 237f.

117 Foucault [1977].

118 Castoriadis [1981; 375f].

119 *Ibid.*, 376 und 377.

120 *Ibid.*, 377.

121 *Ibid.*, 375 und 377.

122 *Ibid.*, 203; Castoriadis bezieht sich selbst ausdrücklich positiv auf Fichte und seine Einsicht in "die grundlegende Rolle der Imagination im radikalen Sinne" im Begriff der "produktiven Einbildungskraft" in der ersten Wissenschaftslehre (251 Fußnote).

123 *Ibid.*, 249f.

124 *Ibid.*, 246.

125 Vgl. *Ibid.*, Fußnote 106.

LITERATUR

Bandrillard J. [1982]. *Der Symbolische Tausch und der Tod*. München.

Costariadis C. [1981]. *Durchs Labyrinth. Seele, Vernunft, Gesellschaft*. Übersetzt von Horst Brühmann. Frankfurt/Main.

Costariadis C. [1984]. *Gesellschaft als imaginäre Institution. Entwurt einer politischen Philosophie*. Übersetzt von Horst Brühmann. Frankfurt/Main.

Foucoult H. [1977]. *Die Ordnung des Diskurses*. Inauguralvorlesung am College de France – 2 Dezember 1970. Frankfurt/Main, Berlin, Wien.

Glaser H. [1979]. *Spießer-Ideologie*. Frankfurt/Main-Berlin-Wien.

Habermas J. [1974]. Können komplexe Gesellschaften eine vernüftige Identität ausbilden? in: Habermas J., Henrich D., *Zwei Reden*. Frankfurt/Main.

Habermas J. [1976]. *Zur Rekonstruktion der Historischen Materialismus*. Frankfurt/Main.

Hegel G. W. F. [1971]. *Werke 1. Frühe Schriften*, Frankfurt/Main.

Honneth A. [1977]. Über Althussers Gesellschaftstheorie Geschichte und Interaktions-verhältnisse. Zur strukturalistischen Deutung des Historischen Materialismus in: Jaeggi U., Honneth A., *Theorien des Historischen Materialismus*. Frankfurt/Main.

Löwenich F. [1987]. Enteignete Sinnlichkeit. Modernes Schillern: ästhetische Erziehung als Ästhetissierung der Realität, in: *Der Deutschunterricht*; Jg. XXXIX, Heft 1, Februar.

Löwenich F. [1988a]. Integration durch Kultur. Adorno mit Lacan, Lacan mit Adorno, Eine Skizee (gekürzte version) in: *Ästhetik & Kommunikation* 67/68, Jahrgang 18, Februar.

Löwenich F. [1988b]. Ästhetisierung der Realität. Das Imaginäre der Kultur: Kulturelle Integration in: *Wo Es War* no. 5/6, Dezember.

Löwenich F. [1990]. *Paradigmenwechsel über die Dialektik der Aufklärung in der kritisch revidierten Kritischen Theorie*. Würzburg.

Luhmann N. [1964]. *Funktionen und Folgen formaler Organisation*. Berlin.

Luhmann N. [1969]. *Legitimation durch Verfahren*. Darmstadt und Nauwied.

Morris Ch. [1973]. *Zeichen, Sprache, Verhalten*. Düsseldorf.

Pfankuch K., Lepper G. [1981]. Persönlichkeit und Masse. Zur Politisierung der Existenz – und Lebensphilosophie Anfang der 30er Jahre in: *Notizbuch 4*. Berlin.

Rossi-Landi F. [1981]. Über einige nach-Manische Probleme in: A. Eschbach (Hg.), *Zeichen über Zeichen. 15 Studien über Ch. W. Morris*. Tübingen.

Poznań Studies in the Philosophy
of the Sciences and the Humanities
1991, Vol. 22, pp. 173–195

Tomasz Maruszewski

EVERYDAY KNOWLEDGE AS REPRESENTATION OF REALITY

1. Introduction

Analyses of everyday knowledge are carried out by philosophy, sociology and psychology. However, definite approaches or ways of analysing of everyday knowledge have become a kind of ritual. For example, it is in good taste to begin a standard handbook on behavioural science by opposing everyday against scientific knowledge. It is widely shown, that everyday knowledge is not by far as good as scientific one both with respect to the validity of explanation, the precision of prediction, internal consistency and cognitive economy. Books by Kerlinger [1964], Nowak [1977], Shaver [1981] or many others may serve as examples.

Such a procedure seems to be understandable both from the philosophical and psychological point of view. Researchers while generating personal constructs that order their own experience (here: scientific experience) start from similarities and differences in strings of events [Kelly, 1955, 1970]. For example, a prerequisite for generating such a construct as "scientific knowledge" is contraposing it to something what is totally different. Most often everyday knowledge has taken the blame. One might just as well to contrapose scientific knowledge to religion, what in fact was the greatest preoccupation of 19-th century positivists [Życiński, 1983] or it might be contrasted with ideology. However, recently in both metascientific and metatheological considerations or in considerations concerning ideology similarities are stressed between scientific knowledge and other conceptualizations of reality. Such an attempt is put forward in Barbour [1983] who stresses the importance of models both in religion and scientific cognition or in Życiński [1983] who indicates theological and methodological consequences of the existence of "black holes". Another example of how these differences become blurred is the phrase "scientific communism".

Reversing to the contraposition of everyday and scientific knowledge mentioned earlier, we may state that seeking for contrasts between them resulted in that differences being seen where they were actually absent. To put the problem in a quite general way we could say that scientific and everyday knowledge may differ in some respects (e.g. methods of verifying of hypotheses) while they may be similar in other. Such similarities take place among cognitive procedures responsible for generating everyday and scientific knowledge — what is called "a context of inquiry" [Maruszewski, 1983].

In the present paper an attempt is made to interpret everyday knowledge as a certain form of representation of reality. However, it will not be restricted to a comparison of everyday and scientific knowledge or to a comparison of everyday representation of reality with representation resulting from scientific cognition. A comprehensive analysis of everyday knowledge calls for recognition of its multi-dimensional character.

2. The concept of representation

Neobehaviouristically oriented analyses of cognitive processes stress that a main feature of these processes is their representational character [Moroz, 1972]. However, Moroz though complaining above the unreflecting nature of many psychological conceptions does not give due consideration to the analysis of the concept of representation. Thus, it is high time to analyse the concept more carefully and precisely. Generally speaking representation means re-presenting something. In Marxist psychology the term "reflection" is often used — though almost the same it is not equal to representation. Intuitively: by presentation of something we mean a valid and precise rendering of relations in the world. Let us make some terminological assumptions in order to analyse the concept of representation more precisely. Let O be the represented object, and O' be a representation of this object. We may treat O as an ordered pair $<A, R>$ where A is a certain set of features and R is a set of relations among these features. O' may be presented as an ordered pair $<B, S>$ where B is a set of features and S is a set of relations among them. In the theory of measurement [Coombs, Daves, Tversky, 1970] an assumption is made that O' is a representation of O when and only when if there exists a function f which to every element $x \in A$ assigns one and only one element $f(x) \in B$ in such a way that if there is a relation M between two elements x and y in set A it will be accompanied by a relation S between elements $f(x)$ and $f(y)$ in a set B. Elements $f(x)$ and $f(y)$ are images of x and y respectively. In this moment arises question whether a representation and a represented object or phenomenon are isomorphic or homomorphic to each other. In the case of isomorphism the function f is equiequivoal; it means that it assigns one element from O one element from O' and *vice versa* — one element from O' corresponds to one element from O. In the case of homomorphism the assignment is many-equivocal, i.e. many objects may have one common representation.

Of the above mentioned interpretations of representation I prefer the latter. Although it suggests that representation is not very precise image of some phenomenon or object, there are some arguments supporting this view:

1. Our sensory organs are characterized by a limited extent of sensitivity. For example, we are not able to perceive certain parts of electromagnetic radiation. Thus, in spite of the fact that the vawelength for UV and infrared radiation is different, we are not able to notice any difference between them, as we cannot see them at all. However we may invalidate this argument by laying that people can distinguish these two kinds of radiation providing they make use of other kinds of representation and not the sensory one. On the other hand if we accept this argument we should also acknowledge that all important dimensions of studied objects or phenomena has already been established, and hence, everything which is not directly available to our sensory organs may be studied and measured in a more indirect way, by means of other measuring instruments or procedures of inference. Yet another assumption seems to be far more justified: the world is more complicated than it is suggested by our perceptual and cognitive analyses. Thus, something being interpreted as equivalent at the moment may, in fact, be differentiated in respect to a criterion which has not yet been conceptualized or registered by our sensory organs. Numerous psychophysical studies have revealed that our ability to differentiate objects is not unlimited. We treat objects as identical which differ from each other less than threshold of difference.

2. One of the most basic functions of the human brain is its ability to reduce variety. There are anatomical and functional evidences supporting this position. For example, the retina of the eye contains about 126 million rods and cones, while the optic nerve composed of

the axons of the ganglion cells contains only about 1 million fibres [Grusser, Grusser-Cornehls, 1978]. Hence, we may conclude that there must be a great reduction in the number of visual information transmitted over 1 million optic nerve fibres. From the functional point of view the importance of the reduction of variety has been documented by many psychological experiments which have shown that all perceptual processes involve a comparison of sensory data with the data stored in memory [Neisser, 1967; Lindsay, Norman, 1972]. Of course, a limited capacity of the short term memory restricts the number of elements that may be compared at a given moment. It must be added that even the simplest act of perception may be considered as an act of categorization. When I perceive object x and say that x is a chair. It means that I have noticed the characteristic features of chairs in object x, but — at the same time — I may ignore whether it is made of wood or steel. Certainly enough, specific traits of an object may be introduced into representation if action undertaken by an individual would require consideration of them. For example, a model being photographed will focus her attention on the colour of the chair and make sure it matches her dress while a clerk will be more interested in whether (or not) the chair is comfortable. As actions undertaken by an individual never require consideration of all features of a perceived object, thus representation of this object is multiequivocal.

3. Biological and economical aspects also support the homomorphic nature of representation. It would not be useful — from the biological point of view — to respond to all features of a particular object. Some features are biologically unimportant, hence the organism does not have to consider them during decision making whether to perform some apetitive or defensive action. Man is relatively independent on biological meaning of objects and thus his or her cognitive processes may be more "disinterested". However, people usually try to reduce their cognitive effort. Empirical studies have shown that striving for cognitive economy makes people carry out their cognitive operations not on the representations of particular objects (in that case we would still speak of isomorphism), but on the so called basic level representations [Rosch, 1978]. These representations occupy an intermediate position within the hierarchy — between general categories and the representations of particular objects. For example, while naming objects in pictures children usually start with names one from the basic level category — they say it is a car, but they never use the term

"vehicle" or "Fiat". Studies on the semantic decision time have also proved that the basic level objects occupy a privileged position.

Summing up, we may say that homomorphic representations are characteristic of many living organisms. The more primitive an organism and the less specialized his sense organs the higher level of homomorphism (i.e. the more different objects have one common representation). In an extreme case the whole variety of the world might be reduced to two categories "good — bad" or "useful — harmful". In fact, man still generates homomorphic representations and — if we accept the above arguments — never achieves isomorphic representation, although asymptotically approaches it.

It must be added that in some cases another kind of assignment has been found, namely the equi-multivocal assignment. Thoroughly studied cases have been delivered by the psychology of perception — they are various ambiguous figures where the same geometric pattern may be interpreted in different ways. Similar phenomena may be observed in the case of representations more complex than perceptual ones, namely in the case of the symbolic representations of a given object or person. Two points, however, should be stressed. Firstly, ambiguous figures or their symbolic analogs are not very typical of human cognitive processes — they usually evoke surprise and astonishment in subjects. Thus if we are interested in the essence of representation, we are allowed to analyse such cases as exceptions to the general rule. Secondly, the nature of ambiguous figures seems to suggest that the same individual may use various mapping rules of real object at different moments. A knowledge of these rules let him or her realize that apparently different representations refer to one and only one object. It presupposes the existence of a "meta-knowledge" by means of which this stability may be specified.

3. Everyday knowledge — an attempt at a psychological and philosophical analysis

Everyday knowledge, though often falling a prey to comparisons with scientific knowledge, has not become an object of deeper scientific analysis. Using a literary trick we may say that what was written so far represents scientists' everyday knowledge of everyday knowledge. As a result, such analyses are of postulative nature and do not result in empirical studies.

However, since the fifties a number of psychological analyses had been published aiming at rehabilitation and even ennoblement of everyday knowledge. The beginning was made by works by Kelly [1955, 1970], although Heider's "The Psychology of Interpersonal Relations" [1958] turned out to be the most influential book. Heider states that psychology does not need investigators "seeking further empirical and experimental facts"[1958, p. 4]; this branch of science needs "conceptual clarification as a prerequisite for efficient experimentation" [ibidem] and that is just what the analysis of everyday language should allow us to do. Heider is of the opinion that psychological knowledge is encoded in everyday language and the principal task of psychologists is to reconstruct all the rules constituting this knowledge [ibidem, p. 4]. Heider does not discuss a problem whether the language itself is a source of this knowledge or only a vehicle for this knowledge resulting from individual collective or social cognitive activity.

Heider's book focused psychologists interest on attributional processes. Studies on these processes were introduced by Jones and Davis [1965] and Kelley [1967] and reached epidemic levels in social psychology of the sixties. Various authors tried to reconstruct everyday knowledge using scientific methods - mainly experimental methods, which assume manipulating some variables and concurrent controlling or isolating another variables. Therefore, it is not surprising that the results of these studies made many authors assume that people creating everyday knowledge apply quite simple cognitive procedures. As a matter of fact it is possible to reduce them to the method of agreement and the method of difference (Mill's canons of induction [Laljee, 1981; Maruszewski, 1983]). As it can be clearly seen the neobehaviouristically oriented early version of attribution theory suggests that ordinary people use cognitive processes that are a little bit simpler than processes used by psychologists themselves (of course, it does not mean that these processes cannot lead to reliable psychological knowledge).

Criticism of these ideas in European psychological and philosophical thought has emphasized the fact that both on the level of the research methods used and results obtained projection of social and cognitive experience of psychologists is displayed. Harré [1981] stresses the ethnocentric bias of the studies on attribution in which mapping of the American way of life in a laboratory experiment have become a rule. One of the symptoms of this ethnocentrism is delivering information, which is to be the basis of attribution, in form of documents such as specially prepared descriptions of other people. "There is a very strong tendency in United States to create a surrogate world of paper, "documents" which for many practical purposes stand for real things and, in particular, for real people. We must ask, then, whether the studies reported are studies of how people handle of personality documents — for example, how people make written attributions of explanatory psychological attributes of others, or deal with others written attributions. A survey of the literature shows quite unequivocally that most literature presented as the "results" was generated from documentary work undertaken by the folk " [Harré, 1981, p. 141]. Subjects in such studies usually responded by creating documents, as well — for example by putting signs on rating scales. Thus, it is not surprising that subjects in these studies used simple cognitive procedures, although many other data suggest that these procedures are either more complicated or at least more diversified. Kruglanski, Hamel, Maides and Schwartz [1978] and Laljee [1981] suggest that attribution processes are of greater complexity and state that the basic cognitive procedure is the hypothetic-deductive method; previously I argued that man uses a certain variety of idealization-concretization procedure [Maruszewski, 1983, 1985]. Let us note that what all these approaches, both classic and "revisionist" have in common is — looking for a frame of reference for the analysis of everyday knowledge and finding it in the analysis of scientific knowledge. The only differences are those which refer to theories within the philosophy of science which for such an analysis is to be based upon - now it is inductionism, now falsificationism, then idealizational theory of science.

Let us try to answer the question what everyday knowledge is. Is the statement that having power demoralize people part of everyday knowledge? Is the ability to assess as bizarre the reply given by a man responding to the invitation "What about a drink?" by saying "Yes, thanks, I'm an alcoholic (the example quoted by Harré [1984]). Do participants of various training groups use this kind of knowledge when they communicate in "psychobabble"? They use such expressions as "I am feeling really mellow", "What space are you in?" [Heelas, Lock, 1981, p. 6].

The first case seems to be obvious, whereas the two others may be given different interpretations. One can say that the ability to evaluate other people's behaviour may be described either as tact or psychological intuition, but it does not have to be everyday knowledge.

The last example referring to psychobabble might seem very controversial but for people who spend a lot of time dealing with "self-development" or with "extending his or her awareness" such expressions might be an attempt to articulate their experiences.

These examples suggest that the concept of "everyday knowledge" as well as the concept of "common sense" (which is similar, but not identical) are natural concepts with fuzzy borders. They change their borders depending on situational context, users' experience or simply the acquisition level of the concept. Trying to determine the meaning of the concept of everyday knowledge relatively precisely and bearing in mind that it is a natural concept, we may go along two routes: firstly, we may try to establish the referential meaning of the concept, that is, we may point to a group of objects, people or phenomena which are termed with the same concept; secondly, we may try to establish the basic features of everyday knowledge, what in turn should make it possible to establish the structural meaning of the concept (that is, a meaning as a sense). Establishing of the latter should also make it possible to locate a given concept within a network of other psychological concepts.

Most of the works published up to now have adopted the first solution. Psychologists, however, have more often make use of the rather unclear concept of "common sense"; it was applied in very a general way, without further qualification — it is no wonder that common sense has been interpreted not only as a knowledge, but also as skills or specific ways of emotional responding. Schwieso [1984] says, that lack of the clear-cut definition of common sense has polarized psychologists' attitude to it — some of authors attempted to dignify common sense knowledge by considering it as a source of cognition, whereas others have argued that it is useless to deal with common sense from the point of view of science. But "...the various disputants have not found the meaning of common sense (perhaps it seemed common sense!)" [ibidem, p. 43] — it means that the concept is clear and does not require deeper consideration. Schwieso distinguishes four basic ways of understanding common sense: "Common sensation (OED1) — an internal sense or consciousness linking the five senses; (OED2c) — ordinary or untutored perception; (OED4) — (philosophical) primary truths forming the grounds of our experiences. Ordinary intelligence (OED2) — distinguishing normal people from mentally ill or handi-capped. Good sense (OED2b) — tacit judgment, practical understand-ing: seemingly quality of specific persons or actions. Common opinion

(OED3) — those judgments, feelings (and beliefs) common to mankind or a community" [Schwieso, 1984, p. 43].

It is difficult to treat the above list as a classification of various meanings of common sense. For example, the element designated as OED4, that is more closely characterized as "...convictions or cognitions which form the ground of our experience and beyond which no appeal to more reliable evidence is possible " [ibidem, p. 44] is hard to differentiate from OED3 that is "...processes which, because they are shared by others, enable us to reach similar conclusions concerning particular situations" [ibidem, p. 44]. If we distinguished them we would have to assume the existence of some universal cognitive processes, independent of socialization. Meanwhile, as shown by Asch's experiments [1951) people are very sensitive to how others perceive even very simple features of objects (eo ipso it would be very hard to speak about any cognitive processes independent of socialization provided that Asch results are not an artifact). It must be added that the description of OED3 does not seem very precise as it now refers to judgments common to mankind (what sort of judgments — mathematical or logical truths? but why call them as common sense judgments and not mathematical or logical ones) and then it refers to judgments shared by a certain community. The latter is more acceptable; it allows us to speak about various common senses in various social groups or communities. Thus, we might conclude, that there is no single common sense but there are various common senses which not overlapping each other, even to the point of being inconsistent. It is obvious that people acknowledge as the truest or most valuable the common sense of their own social group, whereas they consider the other "common senses" as false, wrong or less useful. Thus, if common sense was not a collection of statements as "water is wet", "ice is cold", etc., the existence of various common senses would have to be assumed. Statements constituting OED3 are synthetical sentences. They are not only descriptive but also evaluative (although to a lesser degree than OED2 and OED2b). Psychologists describe OED3 with such concepts as "ideology, attitude, cultural difference, social norm, moral standard, etc." [ibidem, p. 44]. I think that using a collective term such as "everyday experience" is more convenient than using a rather loosely structured list.

Quite a similar proposal — what is interesting that almost simultaneously — was presented by Fletcher [1984]. He distinguishes 3 aspects of common sense. Firstly, he considers common sense "a set of shared fundamental assumptions" [*ibidem*, p. 206] which make it possible to maintain and generate an explicable and reasonable world view. As examples of such assumptions the author mentions assumption concerning the existence of the world independently of our perception, of stability in time of causal relationships, assumption of consciousness characterizing other people, etc. These fundamental assumptions are unanimously accepted by the Western culture and they are presumably universal in character (the last statement is in my opinion, rather dubious); these assumptions are taken for granted and they are neither questioned nor even articulated. Fletcher is of opinion that explanatory power of these assumptions is so great, that they may be considered axioms.

In the second meaning, common sense is interpreted as a set of cultural maxims and beliefs shared by groups of people. In contrast to the previous assumptions these are accepted explicitly and they are relativistic in nature. They may occur in the shape of proverbs or longer fables. Two basic features differentiate them from the first group: a) in a particular culture they are not universally accepted (not to mention differences between cultures); b) they are often articulated or defended by people [Fletcher, 1984].

Finally in the third meaning common sense is understood as a "shared way of thinking". Man acquires this way of thinking within his own culture in an unconscious and automatic way. This lack of conscious access makes it similar to the first way of understanding common sense, where it is defined as a set of fundamental axiomatic assumptions. The difference between the first and the third meaning consists in the fact that these ways of thinking reveal both similarities and differences within a particular culture and between cultures [Fletcher, 1984].

The idea presented above and the idea proposed by Schwieso overlap. What Schwieso calls "(philosophical primary truth forming the grounds of our experiences" is in Fletcher's work included in the first meaning of common sense. "A shared way of thinking" mentioned by Fletcher includes both "ordinary intelligence (OED2)", "good sense (OED2b)" and "ordinary or untutored perception (OED2c) in Schwieso's article (the latter specifies particular cognitive processes, whereas Fletcher traces them within one generic concept). Ultimately, the second meaning of common sense in Fletcher's work corresponds to "common opinion (OED3)" in Schwieso.

It seems that the most general feature that links the views of both authors is the fact that they implicitly accept the assumption of the existence of a difference between everyday knowledge and processes that has led to its generation. Some part of this knowledge is axiomatic in nature and the rest are synthetic statements that may be verified empirically (presumably the axiomatic part is in fact also a set of synthetic statements; however, for want of relevant methods of empirical verification they are accepted somehow *a priori*.

Everyday knowledge is only in part a result of individual cognitive activity. Some part of it is acquired in an unreflexive, automatic way during socialization processes. Socialization pressures may lead to changes of behaviour at first and then to changes within knowledge (change of knowledge would be epiphenomenon of behaviour). Forming a knowledge as a result of behaving incongruently to one's own attitudes may serve as a classical example. The individual usually attempts to justify his or her behaviour: he or she explains it by adopting an attitude that is exactly opposite to the original; ultimately the individual changes his or her primary attitude. Reasoning is based on a desire for saving one's face: "because I am a rational man, I behave in a way that is congruent with my beliefs; I behaved in such and such a way, thus I have a relevant belief that is the best way of behaving in this particular situation" But it is only one of possible examples of how information on rules directing one's own behaviour and also indirectly on rules governing other people's behaviour (providing that other people have been subject to the same socialization pressures) is acquired. Further presentation of situational determinants that change not only people's behaviours but also changes the cognitive representation of oneself and the world, would be beyond the scope of this paper; for a detailed description of these determinants see Aronson [1972].

It should be stated here what everyday knowledge is and how the notion is related to the concept of "common sense" discussed above. As I have already suggested the concept of everyday knowledge is presumably a natural concept and thus it is hardly possible to establish its exact limits. Of course, it does not mean that one should not to attempt to find such a set of features which could be used for characterizing prototypical examples of the concept "everyday knowledge". The first part of the definition contains qualification of superordinate concept, i.e. the concept of knowledge. Although there is no full agreement among psychologists upon what knowledge is, most of them, however, would accept the idea that knowledge is an ordered and systematized set of information about the world and about oneself. The problem arises when we try to point out the generic difference, i.e. when we try to determine what makes a given set of information be considered as everyday knowledge.

Topolski considers everyday knowledge as an accumulation of information about "relationships between facts, more often about co-occurrence of some facts, or traits, which we know during our everyday practice" [Topolski, 1978, p. 44]. It has the shape of inductive generalizations that do not go beyond directly available information. The examples cited by Topolski ("hot ovens burn", "people are envious", "dark clouds announce rain") do not form a homogeneous group. The solution proposed by Topolski consists in determining whether a particular knowledge is common sense knowledge on the basis how this knowledge has been acquired ("inductive generaliza-tions"). Though in my opinion the solution is highly justified, it cannot be accomplished at the present moment. It should be noticed that the concept of everyday practice is not clear. For example, how should the everyday practice of a pupil be interpreted; to a large extent it consists in acquiring various pieces of information — moreover a curriculum is constructed in such a way that it requires transferring information found in one situational context to others contexts (thus, what we get is unjustified generalization). Another example : a worker operating on a highly complicated machine uses his "scientific" or everyday know-ledge? And how should it be accounted for if the worker accidentally adopts a more effective work strategy than specified by operating instructions? A lot of similar questions may be raised.

A further property of everyday thinking mentioned by Topolski [1978], namely, its phenomenalistic character presumably applies only to some cases of such thinking. As I have tried to show elsewhere [Maruszewski, 1980, 1983] a characteristic feature of everyday know-ledge is making use of phenomenalistic procedures. For example, experiments on information integration have revealed that people use additive or averaging models with variable weights; thus it might be concluded that while making judgment people ascribe greater weight to some factors and smaller to other factors. In other words, they recognize some factors as more significant and other as less significant — this cannot be brought into agreement with the phenomenalistic interpretation of thought procedures. Undoubtedly, people use phenomentalistic strategies in some cases but they are also able to use essentialistic procedures. If we acknowledge that in a methodological and psychological analysis both performance and competence should be taken into consideration, the results of the experiments mentioned above might necessitate revising the statement on the phenomenalism of everyday knowledge.

In this place we may agree with Topolski's criticism of the interpretation given by Nikitin [1975] who recognizes conciseness and enthymemathicity as main features of everyday knowledge. In fact, everyday thinking is based on unverbalized presumptions, nevertheless an individual is able to articulate these presumptions in a clear way as well as we are able to construct arguments supporting them. As Fletcher has suggested previously, people tend to completely display all their presumptions of reasoning, while presenting or defending their views (common sense defined as a commonly shared opinion).

Let us then present a positive solution to the problem of basic features of everyday knowledge. It seems that the conclusive feature for a particular knowledge to be considered everyday knowledge is the fact that it is a result of individual practice. Although we have already mentioned that the concept of practice is unclear, it seems that other features are of secondary importance. So far, we have dealt with the concept of practice in terms of common sense and therefore considered it obvious. Now, in order to define practice most accurately, two kinds of it are to be distinguished: "physical practice" which applies to inanimate objects and either changes them or stabilizes their variable properties (e.g. the temperature inside) and "social" practice which applies both to inanimate objects (e.g. any stuff used for artistic purpose) and animate organisms and either modifies them (e.g. upbringing, training of animals) or helps keeping them unchanged (e.g. psychoprophylactic treatment). The restrictions imposed on physical and social practice are different. As concerns physical practice, affecting a certain material is limited by the physical properties of this material; thus people are relatively free to change it. As a result, a wider spectrum of the properties of a given object can be studied. To give an example: a child may study the inner structure of a toy dog while the same procedure on a living animal would require reducing it to a physical object (e.g. by blocking it defense mechanisms). As concerns social practice, the range of possible influence is limited by social norms that prohibit certain behaviour (e.g. the incest taboo); on the other hand, in any interactions other persons' plans, intentions and emotional properties must be taken into account. In other words, social interaction is the resultant of an individual's intentions and that what he or she finds out (by negotiating) would be considered proper by another individual. Some rules governing these negotiations are to be found in Fletcher's first definition of common sense where it is considered identical with "a set of shared fundamental assumptions" [1984]. In the basic rule an assumption is made that there is an essential similarity between interacting individuals. This similarity comprises the ability to plan one own's activity, rational directing one's efforts towards the desired goals, the skill of making capital of information, the ability to perceive one's own mental processes as well as the ability to see the discrepancy between intentions and results. On the other hand, the similarity between mental processes is not that strong — we may call it "generic". It boils down to the fact that interacting individuals function under familiar and acceptable circumstances. If this basic rule of "similarity" is satisfied, an individual can foresee another individual's behaviour (or to express it more carefully, range of possible behaviour) and negotiate the most desired results. It appears that if everyday knowledge is a result of an individual's practice then it must be taken for granted. If it has proved to be efficient then it must be true. Having tested it many times, people tend to accept everyday knowledge as their own. It is not necessarily so. Knowledge which is believed to be a result of one's own conceptual activity might just as well be part of social consciousness. Stereotypes with no empirical grounds may serve as classical examples. An individual may accept a widely shared assumption that the French are gourmets and well-behaved persons even if he or she has never met a Frenchman. And even more, one can be strongly convinced that the statement is true! Where do such beliefs stem from? They may be imposed by a group or a respected individual. Still, it seems that they are adopted by means of an intermediate mechanism that is a by-product of any interaction. In order to obtain the desired goals, people try to make others think they are alike (cf. the similarity rule, p. 183). Only then, the course and result of an interaction can be negotiated. Supporting certain beliefs is the price people pay for making such negotiations possible. Certainly, one can reject widely shared beliefs. This happens if an individual has developed a strong system of own beliefs (he may even adopt the role of an opinion leader and try to persuade others to accept his points of view); or if he does not approve of shared beliefs because they are inconsistent with his set of values. As a result of such conceptual (or even behavioural) nonconformity an individual may be considered deviant.

It must be noticed that the influence of social practice on such subjective properties of everyday knowledge as its obviousness is exercised both directly and indirectly. The direct influence results from reinforcement if certain beliefs are accepted as the acceptance is usually directly rewarded. The indirect influence of social practice is exerted if individuals in interactions hold certain beliefs in order to negotiate or obtain the desired results though they do not approve of these beliefs. As concerns physical practice, it influences everyday knowledge only in a direct way. From the point of view of physical practice, beliefs are confirmed (or not) and an individual considered them obvious depending on how regular and frequent they are. The intermediate mechanism mentioned above cannot be applied here as people do not negotiate with inanimate objects.

As shown above, everyday knowledge is gained by means of multiple mechanisms, the differentiation of these mechanis+s depends on its content. Therefore it would be wrong to assume that there exists a universal mechanism responsible for how this knowledge is formed. Because everyday psychological knowledge and everyday social knowledge fulfill significant self-presentational function and strongly influence the information man is able to obtain in social interaction, hence referring in the analysis of its generation only to rules of cognitive psychology may be misleading. R. Harré [1981] considers such treatments to indicate a dramatic convention typical of psychologists which allow them to act as "pure scientists". The criticism is first of all directed to the attribution theory and brings to light some implicit assumptions put forward by attributionists — yet, it seems the criticisms is too sharp as it identifies psychological knowledge with everyday knowledge. If all sorts of psychological knowledge were to be defined in that way the argument would be of no avail since it is derived from everyday knowledge!

4. Dimensions of everyday knowledge

As already mentioned (p. 172), everyday knowledge is of multi-dimensional character. The dimensions of everyday knowledge are of no equal importance, in the present analysis we will start with those rated

188

highest and proceed to those rated lower; we will attempt to focus on the most important dimensions therefore the presentation will be incomplete.

THE PRAGMATIC DIMENSION. From the subjective and objective viewpoint, it is the most important dimension. Subjectively, an individual is intent on acquiring a knowledge which guarantee efficient functioning under various conditions. From the objective viewpoint, we may refer to the theory of learning and point out that only behaviour (and cognitive structures responsible for it) leading to reinforcement is strengthened. Reaching goals by an individual may serve as an example of such reinforcement. The question whether efficient behavioural techniques are based on genuine knowledge is of secondary importance. Let us consider Topolski's [1978] example: "people are envious". In a particular culture (here: in a competitive rather than co-operative culture) such an opinion justifies concealing one's own achievements and manifesting modesty or even indigence. Such sociotechnical procedures bring advantage and do not make rivals feel envy; consequently, their vigilance is put to sleep. It does not matter if people are really envious or of their envy has been properly accounted for. In addition, as it can be clearly observed, the pragmatism is individual, i.e. an individual approves of judgments which are efficient to him and not necessarily to those which guarantee successful functioning to other people. It should be noticed that if the above-mentioned judgment is accepted by a large number of people it cannot guarantee successful functioning since everybody realizes the conventional nature of numerous statements and ways of behaving. The situation resembles "the prisoner's dilemma" — deciding on individualistic strategies by all the participants leads to worse results than if they co-operated.

Assigning the dominant importance to pragmatism may be criticized as there are evidences that everyday knowledge is inefficient or, at least, less efficient than other kinds of knowledge. People behave irrationally and they do not make use of all information available or overrate the best accessible information. These facts cannot be denied. The proposition of pragmatism is upheld from the essentialistic point of view [Cohen, 1981]; according to it the pragmatism of everyday knowledge is tied up with its essence whereas counter-evidences merely manifest its essence. Such manifestations are influenced by multiple side-effects — as for example reduced possibility of information processing. Let us carry out a simple experiment to prove the pragmatic nature of everyday knowledge. How could the behaviour of an individual be accounted for if his or her knowledge was not efficient? How could his or her efficient functioning be explained? In terms of emotional processes or fortuitousness? Does tis world show courtesy and eagerly submit itself to people's intentions? Such statements are obviously absurd. Thus, it can be taken for granted that as a matter of course everyday knowledge is directed to the efficient realization of an individual's intentions or plans. Whenever everyday knowledge does not satisfy the above condition, side-effects are to blame.

The question whether everyday knowledge is pragmatic is of relative character. The level of pragmatism is conditioned by social factors and people's individual properties. Under given circumstances an individual need not try to obtain the best possible results but he or she will aim at the results which fall within personally and socially acceptable limits. For example, the pragmatic knowledge of a little child buying something by weight differs from that of adults as a child cannot calculate the right price. The pragmatism of a child results from an assumption that the shop assistant knows the price and demands the right charge. As it seems, a child's pragmatism is widely accepted and though children could be easily cheated it is rarely a case. As a child acquires the ability to calculate and judge it modifies the basis for its pragmatic reasoning. The example reveals another interesting feature of the problem – the level of pragmatism is adjusted to that of a partner in interaction. The shop assistant realizes what a child is capable of and he know it considers him a competent person; therefore he behaves according to its expectations. Cheating children, though easier than cheating adults, usually evokes social disapproval as children cannot be regarded as coordinate partners in social interactions.

The pragmatic aspect of everyday knowledge has been strongly emphasized by Kruglanski [1980]. He claims that a cognitively active individual displays *teleological functionalism*, i.e. he or she aims at solving such problems which enable him or her to obtain specific external or internal goals.

THE "VERACITY" DIMENSION. The dimension reflects a tendency that judgments constituting everyday knowledge correspond to reality. It is assumed that everyday knowledge refers to the classical definition of truth. As scientists, we must be aware of the methodological and philosophical difficulties which are tied up with the concept; nevertheless, it suits our purpose. The major difficulty stems from the fact that in order to evaluate whether a judgment corresponds to reality we need independent information on what this reality is like. Consequently, a question arises whether this information itself is a kind of judgment and how it corresponds to reality; the procedure seems never-ending. Everyday knowledge is characterized by its realism, it is assumed that all its statements refer to reality and it is possible to determine whether or not they are consistent with reality. One fails to notice the difficulties in determining the degree of this correspondence; the veracity of a judgment is assessed either subjectively or it is based on good authority (this is often misused in mass media where certain judgments are made more reliable by ascribing them to experts).

It should be emphasized that pursuing veracity is far more differentiated in everyday knowledge than in science. Science employs several procedures of epistemological impoverishment — i.e. different sorts of idealization [Nowak, 1977, 1980]. In such procedures only the most important and essential factors are taken into account. As a result, the image of reality becomes impoverished but at the same time a precise analysis of an individual's behaviour is made possible. The level of impoverishment may be diminished by gradual inclusion of less important factors into analysis but not all the determinants can be reconstructed. Theoretically, several uncontrolled, random factors lie outside the cognitive abilities of a professional researcher, so do some side factors of minor importance.

As concerns everyday knowledge, the procedures of epistemological impoverishment — paraidealization, which I have analysed in details elsewhere [Maruszewski, 1983] — are accompanied by the procedure of epistemological enrichment [Nowak, 1985]. "The point of departure is a real object with all its properties and the task is not to mentally take some of them away but, on the contrary, to add some new ones. There comes into being a mental construct which have more properties than the prototype. The construct and the abstract differ, the construct being enriched and not impoverished in comparison with the real object [...]. What is essential, the procedures of epistemological impoverishment and enrichment seem to be the coordinate factors of everyday knowledge" [ibidem, p. 288]. Nowak suggests schematization as en example of epistemological enrichment: "Schematization is at least a coordinate procedure [...] — the Poles, as other nations, "arrogate" to themselves a lot (e.g. they have an exaggerated image of their own prowess and chivalry)" [*ibidem*, p. 287].

It is only right to agree that epistemological impoverishment is not the only procedure typical of everyday knowledge. Schematization is an essential factor of this knowledge but the point of controversy is whether it should be considered a case of epistemological impoverishment. Let us interpret Nowak's suggestion in terms of the idealizational theory of science. Two interpretations of the concept may be given. Firstly, epistemological enrichment may be viewed as introducing new factors into the picture of the essential structure though, actually, these factors are not to be found within the structure. If chivalry was not a typical trait of the Poles but merely introduced into their "national character" then we would be only right to consider it epistemological enrichment. However, this would raise doubts whether the "veracity" dimension is an important dimension of everyday knowledge. If epistemological enrichment is to be understood in this way I cannot accept it to be an important factor of everyday knowledge. Secondly, epistemological enrichment may be interpreted as moving a factor within the hierarchically organized image of the essential structure. To use the same example: chivalry is in fact an unimportant trait of the Poles but due to schematization it has been moved up to a higher rank. If this interpretation was accepted then the procedure of epistemological enrichment would acquire a relative character. Furthermore, the approach to everyday knowledge which guarantees its correspondence to reality would not be called in question.

A separate issue would be an attempt to establish why certain factors are moved up to higher ranks in the picture of the essential structure. We would have to refer to specific psychological hypotheses. There are several causes for why greater importance is attributed to some factors: e.g., the salience of traits, accessibility within the cognitive structure of an individual, cognitive and motivational consequences (the Poles would rather attribute chivalry than alcohol abuse to themselves). A detailed presentation of these factors lies outside the scope of this paper.

THE INTERPERSONAL DIMENSION. From the traditional viewpoint, everyday knowledge, being the product of an individual's cognitive activity, is individualized in comparison with other kinds of knowledge (e.g. scientific). However, in the light of what we have said so far, the above statement raises doubts. The fact that the basic dimension of everyday knowledge is the pragmatic dimension leads us to conclusion that everyday knowledge is not to be treated as unique. As I see it, the interpersonal dimension of everyday knowledge is the degree to which other people share it. Some judgments and opinions are common to all members of a society, some are shared by particular social groups and rejected by other [cf. Schwieso and Fletcher] and, finally, some are typical of individuals. Since everyday knowledge is pragmatic in the first place its particular elements differ as to their usefulness in communication and social interactions. Universally accepted judgments are of primary importance: they are easy accessible (though not necessarily openly expressed), they may be used as the basis for cooperation, they are easy to interpret. Using such judgments and beliefs makes interaction run smooth and enables negotiating and determining aims. At the same time, the result of an interaction is, so to say, determined beforehand; interactions of that kind are often of ritual character. On the other hand, unique beliefs prolong interactions and necessitate long negotiations; messages are to be plain, still, they often need to be repeated. Let us consider how absorbing it is for a therapist to establish contact with a neurotic patient who, being egocentric, assumes nobody can understand his problems The results of such interactions can hardly be foreseen as they depend on the results of negotiations.

The last, and at the same time, the most frequent case occurs when part of interacting partners' knowledge overlap and part of it is unique. The more their knowledge overlap the more interaction resembles the first case discussed above. The more their knowledge differ the more difficult the communication becomes and the more the result of interaction depends on negotiation.

The great importance of the interpersonal dimension can be exemplified by the difficulties a person encounters while facing a new culture — unwritten requirements and undecipherable social messages make him or her feel inadequate and unable to choose the right form of behaviour.

There is another issue connected with the interpersonal dimension of everyday knowledge — information concerning other people are processed in a different way than that concerning the physical world. As human behaviour is a stream of changes, everyday knowledge processes information in such a way that this dynamic nature of stimulus events is taken into account [Ostrom, 1984]. It is a misunderstanding to speak of "another person's image" since he or she appears to be a string of events. Presumably, a great part of everyday knowledge is registered by means of scripts or cognitive schemata [Abelson, 1981]. Part of everyday knowledge which does not refer to other people or the social world may be registered by means of images or cognitive structures. Since social and physical objects are characterized by a different level of dynamism, changes within these objects may be interpreted in different ways [Ostrom, 1984]. Changes within physical objects are ascribed to outer forces while changes within social objects are attributed to inner forces. These inner forces are intentions and projects to be performed by an individual. Since there is fundamental similarity between people within a given culture, the direction of these inner changes are far easier to establish.

THE ETHICAL DIMENSION. As a matter of fact, the ethical dimension is of minor importance because it is somehow entangled in those three previously described. In some cases (inner conflicts) decision making as to aims for an activity the ethical dimension gains importance. For lack of place it will be not discussed in details here.

***

The above analysis may have not satisfied the readers' expectations. In cognitive psychology the term representation is associated with the mapping of reality in an individual's mind. An analysis of memory structure (both semantic and episodic), information processing responsible for establishing everyday knowledge, introducing and deleting elements seem to be a logical choice for further dealing with the problem of representation. Yet, to accept this point of view would be tantamount to an assumption that formal relationships are of greater importance in the analysis of an individual's cognitive behaviour. In other words, it would be assumed that a student processes information on a professor's likes and dislikes in the same way as he processes information gained from a lecture given by that professor. There are a lot of evidences that both kinds of information are processed differently. That is why our analysis is based on the content elements of everyday knowledge and not, e.g., on the problem of semantic memory. However, it does not mean that everyday knowledge should be entirely

194

analysed in terms of its content. It is for further research to find out how everyday knowledge is connected with, e.g. semantic and episodic memory. If such an analysis is undertaken it will help to improve the scheme suggested by Schwieso [1984], Furnham [1983] and Fletcher [1984].

Department of Psychology
A. Mickiewicz Uniwersity,
Szamarzewskiego 89,
60—568 Poznań, Poland

REFERENCES

Abelson R. P., [1981]. Psychological status of the script concept. *American Psychologist*, 36: 715-729.

Aronson E., [1972]. *Social Animal*. San Francisco: W. H. Freeman & Company.

Asch S. E., [1951]. Effects of group pressure upon modification and distortion of judgment. In: Guetzkow H. (Ed.), *Groups, Leadership and Men*. Pittsburgh: Carnegie.

Barbour I. G., [1976]. *Myths, Models and Paradigms*. New York: Harper & Row Publishers.

Cohen L. J., [1981]. Can human irrationality be experimentally demonstrated. *The Behavioral and Brain Sciences*. 4: 317-370.

Coombs C. H., Daves R.M., Tversky A., [1970]. *Mathematical Psychology*. Englewood Cliffs: Prentice Hall.

Fletcher G. J. O., [1984]. Psychology and common sense. *American Psychologist*, 39: 203-213.

Furnham A., [1983]. Social psychology as common sense. *Bulletin of the British Psychological Society*, 36: 105-109.

Grusser O. J., Grusser-Cornehls V., [1978]. Physiology of vision. In: Schmidt R. F. (Ed.), *Fundamentals of Sensory Physiology*. New York: Springer.

Harre R., [1981]. Expressive aspects of descriptions of others. In: Antaki C. (Ed.), *The Psychology of Ordinary Explanations of Social Behaviour*. London: Academic Press.

Harre R., [1984]. Social elements as mind. *British Journal of Medical Psychology*, 57: 127-135.

Heelas P., Lock A., [1981]. *Indigenous Psychologies*. London: Academic Press.

Heider F., [1958]. *The Psychology of Interpersonal Relations*. New York: Wiley.

Jones E. E., Davis K. E., [1965]. From acts to dispositions. In: Berkowitz L. (Ed.), *Advances in Experimental Social Psychology*. New York: Academic Press (vol. 2).

Kelley H. H., [1967]. Atribution theory in social psychology. In: Levine D. (Ed.), *Nebraska Symposium on Motivation*. Lincoln: University of Nebraska Press.

Kelly G. A., [1955]. *The Psychology of Personal Constructs*. New York: Norton.

Kelly G. A., [1970]. A brief introduction to personal construct theory. In: Bannister D. (Ed.), *Perspectives in Personal Construct Theory*, London: Academic Press.

195

Kerlinger F. N., [1964]. *Foundations of Behavioral Research*. New York: Holt, Rinehart & Winston.

Kruglanski A. W., [1980]. Lay epistemo-logic – process and contents. *Psychological Review*, 87: 70-87.

Kruglanski A. W., Hamel I. Z., Maides S. A., Schwartz J. M., [1978]. Attribution theory as special case of lay epistemology. In: Harvey J. H., Ickes W., Kidd R. F. (Ed.), *New Directions in Attribution Research*. Hillsdale: Lawrence Erlbaum (vol. 2).

Laljee M., [1981]. Attribution theory and the analysis of explanations. In: Antaki C. (Ed.), *The Psychology of Ordinary Explanations of Social Behaviour*. London: Academic Press.

Lindsay P. H., Norman D. A., [1972]. *Human Information Processing Processing*. New York: Academic Press.

Maruszewski T., [1983]. *Analiza procesów poznawczych jednostki w świetle idealizacyjnej teorii nauki* (An Analysis of the Cognitive Processes of an Individual in the Light of the Idealizational Theory of Science). Poznań: Wydawnictwo Naukowe UAM.

Maruszewski T., [1980]. Procesy poznawcze jednostki a idealizacja (Cognitive Processes of an Individual and Idealization) In: Brzeziński J. (Ed.), *Poznańskie Studia z Filozofii Nauki* (vol. 5). Poznań: PWN.

Maruszewski T., [1985]. Are the idealizational procedures used within the scope of common sense knowledge? In: Brzeziński J. (Ed.), *Consciousness: Methodological and Psychological Approaches*. Amsterdam: Rodopi.

Moroz M., [1972]. The concept of cognition in contemporary psychology. In: Royce J. R., Rozeboom W. W., (Eds.), *The Psychology of Knowing*. New York: Gordon and Breach.

Neisser U., [1967]. *Cognitive Psychology*. New York: Appleton-Century-Crofts.

Nikitin E. P., [1975]. *Wyjaśnianie jako funkcja nauki* (Explanation as a function of science). Warszawa: PWN.

Nowak L., [1977]. *Wstęp do idealizacyjnej teorii nauki* (Introduction to the Idealizational Theory of Science). Warszawa: PWN.

Nowak L., [1980]. *The Structure of Idealization*. Dordrecht: Reidel.

Nowak L., [1985]. Recenzja pracy T. Maruszewskiego „Analiza procesów poznawczych..." (Review of a book by T. Maruszewski „An Analysis of Cognitive Processes..."). *Przegląd Psychologiczny*, XXVIII, 285-289.

Ostrom T., [1984]. The sovereignty of social cognition. In: Wyer R. S., Srull T. K. (Eds), *Handbook of Social Cognition*. Hillsdale: Lawrence Erlbaum.

Rosch E., [1978]. Principles of categorization. In: Rosch E., Lloyd B. B. (Eds), *Cognition and Categorization*. Hillsdale: Lawrence Erlbaum.

Schwieso J., [1984]. What is common to common sense. *Bulletin of the British Psychological Society*. 37: 43-45.

Shaver K. G., [1981]. *Principles of social Psychology*. Cambridge: Winthrop.

Topolski J., [1978]. *Rozumienie historii* (Understanding of History). Warszawa: PWN.

Życiński J., [1983]. *Język i metoda* (Language and Method). Kraków: Znak.

Życiński J., [1984]. Posłowie (Afterword). In: Barbour I. G., *Mity, modele, paradygmaty* (Myths, Models and Paradigms). Kraków: Znak.

ELEMENTA

Schriften zur Philosophie und ihrer Problemgeschichte

Herausgegeben von
Rudolph Berlinger und Wiebke Schrader

Band 11: Hfl. 45,—
Djurić, Mihailo: Mythos, Wissenschaft, Ideologie. Ein Problemaufriss. Amsterdam 1979. 219 pp.

Band 12: Hfl. 40,—
Ettelt, Wilhelm: Die Erkenntniskritik des Positivismus und die Möglichkeit der Metaphysik. Amsterdam 1979. 171 pp.

Band 13: Hfl. 30,—
Lowry, James M.P.: The Logical Principles of Proclus' ΣΤΟΙΧΕΙΩΣΙΣ ΘΕΟΛΟΓΙΚΗ as Systematic Ground of the Cosmos. Amsterdam 1980. XIV,118 pp.

Band 14: Sold out
Berlinger, R.: Philosophie als Weltwissenschaft. Vermischte Schriften Band II. Amsterdam/Hildesheim 1980. X,240 pp.

Band 15: Hfl. 90,—
Helleman-Elgersma, W.: Soul-Sisters. A Commentary on Enneads IV 3 (27), 1-8 of Plotinus. Amsterdam 1980. 485 pp.

Band 16: Hfl. 30,—
Polakow, Avron: Tense and Performance. An Essay on the Uses of Tensed and Tenseless Language. Amsterdam 1981. 153 pp.

Band 17: Hfl. 25,—
Lang, Dieter: Wertung und Erkenntnis. Untersuchungen zu Axel Hägerströms Moraltheorie. Amsterdam 1981. 113 pp.

Band 18: Hfl. 30,—
Kang, Yung-Kye: Prinzip und Methode in der Philosophie Wonhyos. Amsterdam/Hildesheim 1981. 143 pp.

Band 19: Hfl. 40,—
Oesch, Martin: Das Handlungsproblem. Ein systemgeschichtlicher Beitrag zur ersten Wissenschaftslehre Fichtes. Amsterdam/Hildesheim 1981. 203 pp.

Band 20: Hfl. 60,—
Echeverria, Edward J.: Criticism and Commitment. Major Themes in contemporary 'post-critical' philosophy. Amsterdam/Hildesheim 1981. 274 pp.

Band 21: Hfl. 30,—
Thomas Hobbes: His View of Man. Proceedings of the Hobbes symposium at the International School of Philosophy in the Netherlands (Leusden, september 1979). Edited by J.G. van der Bend. Amsterdam 1982. 155 pp.

Band 22: Hfl. 30,—
Träger, Franz: Herbarts Realistisches Denken. Ein Aufriß. Amsterdam/Wrzburg 1982. X,139 pp.

Band 23: Hfl. 40,—
Takeda, Sueo: Die subjektive Wahrheit und die Ausnahme-Existenz. Ein Problem zwischen Philosophie und Theologie. Amsterdam/Würzburg 1982. 190 pp.

Band 24: Hfl. 35,—
Mager, Kurt: Philosophie als Funktion. Studien zu Diltheys Schrift "Das Wesen der Philosophie". Amsterdam/Würzburg 1982. 179 pp.

Band 25: Hfl. 50,—
Heinz, Marion: Zeitlichkeit und Temporalität. Die Konstitution der Existenz und die Grundlegung einer Temporalen Ontologie im Frühwerk Martin Heideggers. Amsterdam/Würzburg 1982. 233 pp.

Band 26: Hfl. 50,—
Punter, David: Blake, Hegel and Dialectic. Amsterdam 1982. 268 pp.

Band 27: Hfl. 35,—
McAlister, Linda: The Development of Franz Brentano's Ethics. Amsterdam/Würzburg 1982. 171 pp.

Band 28: Hfl. 60,—
Pleines, Jürgen-Eckardt: Praxis und Vernunft. Zum Begriff praktischer Urteilskraft. Amsterdam/Würzburg 1983. 275 pp.

Band 29: Hfl. 50,—
Shusterman, Richard: The Object of Literary Criticism. Amsterdam/Würzburg 1984. 237 pp.

Band 30: Hfl. 40,—
Volkmann-Schluck, Karl-Heinz: Von der Wahrheit der Dichtung. Interpretationen: Plato; Aristoteles; Shakespeare; Schiller; Novalis; Wagner; Nietzsche; Kafka. Hrsg. von Wolfgang Janke und Raymund Weyers. Amsterdam/Würzburg 1984. 206 pp.

Band 31: Hfl. 40,—
Decher, Friedhelm: Wille zum Leben — Wille zur Macht. Eine Untersuchung zu Schopenhauer und Nietzsche. Amsterdam/Würzburg 1984. 195 pp.

Band 32: Hfl. 30,—
Weppen, Wolfgang von der: Die existentielle Situation und die Rede. Untersuchungen zu Logik und Sprache in der existentiellen Hermeneutik von Hans Lipps. Amsterdam/Würzburg 1984. 146 pp.

Band 33: Hfl. 40,—
Wolzogen, Christoph von: Die autonome Relation. Zum Problem der Beziehung im Spätwerk Paul Natorps. Ein Beitrag zur Geschichte der Theorien der Relation. Amsterdam/Würzburg 1984. 182 pp.

Band 34: Hfl. 50,—
Mitias, Michael H.: Moral Foundation of the State in Hegel's "Philosophy of Right": Anatomy of an Argument. Amsterdam/Würzburg 1984. 197 pp.

Band 35: Hfl. 50,—
Seidl, Horst: Beiträge zu Aristoteles' Erkenntnislehre und Metaphysik. Amsterdam/Würzburg 1984. 214 pp.

Band 36: Hfl. 30,—
Richter, Leonhard G.: Hegels begreifende Naturbetrachtung als Versöhnung

der Spekulation mit der Erfahrung. Amsterdam/Würzburg 1985. 127 pp.
Band 37: Hfl. 35,—
Löbl, Rudolf: Die Relation in der Philosophie der Stoiker. Amsterdam/ Würzburg 1986. 150 pp.
Band 38: Hfl. 70,—
Dempf, Alois: Metaphysik. Versuch einer problemgeschichtlichen Synthese. In Zusammenarbeit mit Christa Dempf-Dulckeit. Amsterdam 1986. 332 pp.
Band 39: Hfl. 80,—
Classen, Carl Joachim: Ansätze. Beiträge zum Verständnis der frühgriechischen Philosophie. Amsterdam 1986. 288 pp.
Band 40: Hfl. 25,—
Middendorf, Heinrich: Phänomenologie der Hoffnung. Amsterdam/Würzburg 1985. 99 pp.
Band 41: Hfl. 80,—
Glouberman, M.: Descartes: The Probable and the Certain. Amsterdam 1986. 374 pp.
Band 42: Hfl. 30,—
Creativity in Art, Religion, and Culture. Edited by Michael H. Mitias. Amsterdam/Würzburg 1985. 134 pp.
Band 43: Hfl. 40,—
Böhm, Peter: Theodor Lessings Versuch einer erkenntnistheoretischen Grundlegung von Welt. Ein kritischer Beitrag zur Aporetik der Lebensphilosophie. Amsterdam 1986. 127 pp.
Band 44: Hfl. 85,—
Weier, Winfried: Phänomene und Bilder des Menschseins. Grundlegung einer dimensionalen Anthropologie. Amsterdam 1986. 337 pp.
Band 45: Hfl. 50,—
Text, Literature, and Aesthetics in Honor of Monroe C. Beardsley. Edited by Lars Aagaard-Mogensen & Luk De Vos. Amsterdam 1986. 229 pp.
Band 46: Hfl. 48,—
Hager, Fritz-Peter: Gott und das Böse im antiken Platonismus. Amsterdam/Würzburg 1987. 165 pp.
Band 47: Hfl. 120,—
Hartmann, Klaus: Studies in Foundational Philosophy. Amsterdam/Würzburg 1988. 446 pp.
Band 48: Broschiert Hfl. 38,—
Gebunden Hfl. 150,—
Berlinger, Rudolph: Die Weltnatur des Menschen. Morphopoietische Metaphysik "Grundlegungsfragen". Amsterdam/Würzburg 1988. 388 pp.
Band 49: Hfl. 45,—
Goedert, Georges: Nietzsche der Überwinder Schopenhauers und des Mitleids. Amsterdam/Würzburg 1988. 168 pp.

Band 50:   Hfl. 60,—
Aesthetic Quality and Aesthetic Experience. Edited by Michael H. Mitias.
Amsterdam 1988. 176 pp.

Band 51:   Hfl. 48,—
Michael H. Mitias: What Makes an Experience Aesthetic? Amsterdam 1988.
154 pp.
154 pp.

Band 52:   Hfl. 27,—
Reinhard Platzek: Zum Problem der Zeit und Zeitbestimmtheit im musikali-
schen Tempo. Amsterdam 1989. 94 pp.

Band 53:   Hfl. 60,—
Patrick L. Bourgeois/Frank Schalow: Traces of Understanding: A Profile of
Heidegger's and Ricoeur's Hermeneutics. Amsterdam 1990. VI, 186 pp.

Band 52   Hfl. 75,—
Thomas Ludolf Meyer: Das Problem eines höchsten Grundsatzes der
Philosophie bei Jacob Sigismund Beck. Amsterdam/Atlanta, GA 1991.
257 pp.

FICHTE-STUDIEN

Beiträge zur Geschichte und Systematik der Transzendentalphilosophie
Im Auftrage der Fichte-Gesellschaft herausgegeben von Klaus Hammacher
(Aachen), Richard Schottky (Wuppertal) und Wolfgang H. Schrader
(Siegen), in Zusammenarbeit mit Daniel Breazeale (Lexington, Kentucky),
Erich Fuchs (München), Helmut Girndt (Duisburg), Marco Ivaldo (Neapel),
Wolfgang Janke (Wuppertal), Reinhard Lauth (München), Kunihiko Na-
gasawa (Kyoto), Faustino Oncina (Valencia), Marek J. Siemek (Warschau)
und Xavier Tilliette (Paris). Die 'Fichte-Studien" wollen die wissenschaft-
liche Erforschung des Werkes von Johann Gottlieb Fichte fördern. Sie
eröffnen Forschern, welche den transzendentalen Gedanken und System-
entwurf philosophisch erörtern, unangesehen der Schulposition und Lehr-
meinung eine Publikationsmöglichkeit. Dabei sollen die historischen Vor-
aussetzungen und zeitgeschichtlichen Kontroversen ebenso zu tieferer Klar-
heit gebracht werden wie die systematischen Konsequenzen heute.
Die einzelnen Bände bringen teils thematisch verbundene, teils vermischte
Beiträge. Außerdem enthalten sie einen Dokumentations- und Rezensions-
teil.
Der Preis für das Abonnement beträgt Hfl. 75,—/DM. 70,—/US-$ 37.50,
zuzüglich Versandkosten. Und für individuelle Abonnenten, die direkt beim
Verlag bestellen Hfl. 37,50/DM 35,—/US-$ 18.75 inkl. Versandkosten.

Fichte-Studien Band 1

Inhalt: *Abhandlungen* Karen Gloy (Luzern), Das Problem des Selbstbewußt-
seins bei Fichte. —Johannes Römelt (München), "Merke auf dich selbst".
Das Verhältnis des Philosophen zu seinem Gegenstand nach dem Versuch
einer neuen Darstellung der Wissenschaftslehre, 1797/98. — Shukey
Kumamoto (Hiroshima), Die Transzendentale Freiheit bei Fichte. — Wolfgang Janke (Wuppertal), Limitative Dialektik. Überlegungen im Anschluß
an die Methodenreflexion in Fichtes "Grundlage" 1794/95 (GA I, 2, 283-
85). — Klaus Hammacher (Aachen), Fichtes praxologische Dialektik. —
Helmut Girndt (Duisburg), Die fünffache Sicht der Natur im Denken
Fichtes. — Wilhelm Metz (Siegen), Die "Weltgeschichte" beim späten
Fichte. — Reinhard Lauth (München), Transzendentale Basis, Materialismus und Religion. — Felix Krämer (Aachen), Maimons Versuch über
Transzendentalphilosophie. Eine interpretierende Skizze der Grundgedan-
ken. — Marek J. Siemek (Warschau), Husserl und das Erbe der Transzendentalphilosophie. — Therese Pentzopoulou Valalas (Thessaloniki), Fichte
et Husserl à la recherche de l'intentionalité. —Wolfgang H. Schrader
(Siegen), Fichte und das postmoderne Denken. —

USA/Canada: Editions Rodopi, 233 Peachtree Street, N.E., Suite 404, Atlanta, Ga. 30303-
1504, Telephone (404) 523-1964, only USA 1-800-225-3998, Fax (404) — 522-7116
And Others: Editions Rodopi B.V., Keizersgracht 302-304, 1016 EX Amsterdam,
Telephone (020) —22.75.07, Fax (020) — 38.09.48

KOSMOPOLITISMUS UND NATIONALIDEE

Amsterdam/Atlanta, GA 1990. 244 pp. (Fichte-Studien Band 2)
ISBN: 90-5183-235-4   Hfl. 75,—/US-$ 37.50

Inhalt: Ives Radrizzani: Ist Fichtes Modell des Kosmopolitismus pluralistisch? Karl Hahn: Die Idee der Nation als Implikat der Interpersonalitäts- und Geschichtstheorie. Wolfgang H. Schrader: Nation, Weltbürgertum und Synthesis der Geisterwelt. Jochem Hennigfeld: Fichte und Humboldt — Zur Frage der Nationalsprache. Johannes Heinrichs: Nationalsprache und Sprachnation. Zur Gegenwartsbedeutung von Fichtes *Reden an die deutsche Nation*. Peter L. Oesterreich: Politische Philosophie oder Demagogie? Zur rhetorischen Metakritik von Fichtes *Reden an die deutsche Nation*. Alois K. Soller: Nationale Erziehung und sittliche Bestimmung. Richard Schottky: Fichtes Nationalstaatsgedanke auf der Grundlage unveröffentlichter Manuskripte von 1807. Klaus Hammacher: Fichte und die Freimaurerei. DOKU-MENTATION: Erich Fuchs: Fichtes Stellung zum Judentum. Erich Fuchs: Fichtes Einfluß auf seine Studenten in Berlin zum Beginn der Befreiungs-kriege. Michel Espagne: Die Rezeption der politischen Philosophie Fichtes in Frankreich.

ETHIK — INTERPERSONALITÄT — POLITIK

Amsterdam/Atlanta, GA 1991. ca. 250 pp. (Fichte-Studien Band 3)
ISBN: 90-5183-236-2   ca. Hfl. 75,—/US-$ 37.50

Inhalt: Marco Ivaldo: Das Problem des Bösen bei Fichte. Alain Perrinjaquet: Individuum und Gemeinschaft in der Wissenschaftslehre zwischen 1796 und 1800. Edith Düsing: Über das Individuations-Problem bei Fichte bis 1800. Villacanas: Fichtes moderner Begriff der charismatischen Legitimität. Carla de Pascale: Nation oder Weltbürgertum? A. Iacovacci: Über Reinhold und Jacobi. Akira Omine: Intellektuelle Anschauung und Mystik. Klaus Hammacher: Freimaurerei, Weltbürgertum und Nation in Fichtes Republik der Deutschen. Yasuhiro Kumamoto: Die Funktion des Verhältnisses von Herr und Knecht in der Philosophie Hegels. DISKUSSION: Georg Geismann: Der Nationalgedanke in Fichtes Rechtsphilosophie. Richard Schottky: Replik auf Geissmann. DOKUMENTATION: Lawatsch: Fichte und die Freimaurerei.

USA/Canada: Editions Rodopi, 233 Peachtree Street, N.E., Suite 404, Atlanta, Ga. 30303-1504, Telephone (404) 523-1964, only USA 1-800-225-3998, Fax (404) —522-7116 And Others: Editions Rodopi B.V., Keizersgracht 302-304, 1016 EX Amsterdam, The Netherlands. Telephone (020) — 622.75.07, Fax (020) — 638.09.48

JOSEF PANETH

VITA NUOVA. Ein Gelehrtenleben zwischen Nietzsche und Freud. Autobiographie — Briefe — Essais. Mit Einleitung und Kommentar herausgegeben von Wilhelm W. Hemecker. Amsterdam/Atlanta GA 1991. ca.
200 pp. (Studien zur Österreichischen Philosophie 17)
ISBN: 90-5183-227-3   ca. Hfl. 60,—/US-$ 30.—

Josef Paneth (1857-1890), Anatom und Physiologe in Wien, war einer der
engsten Freunde von Sigmund Freud und intimer Gesprächspartner Fried-
rich Nietzsches — und damit auch das wichtigste, bislang unbeachtete
"missing link" zwischen beiden. Der Nachlaß Paneths, der hier zum
erstenmal in einer repräsentativen Auswahl veröffentlicht wird, bietet für die
Freud- und Nietzscheforschung gleichermaßen reiches neues Quellenma-
terial. Die Biographie Freuds wird — besonders im Hinblick auf die sonst
mager dokumentierten frühen Jahre — durch Einsichten in den größeren
soziokulturellen Hintergrund, die geistige Situation an der Universität Wien
während der gemeinsam verbrachten Studienjahre, ihre —noch nicht genügend durchleuchtete — philosophische und naturwissenschaftliche Aus-
bildung sowie durch charakteristische Portraits gemeinsamer Freunde, des
von Freud verehrten "väterlich fürsorgenden" Religionslehrers Samuel
Hammerschlag, von Josef Breuer und einigen der berühmtesten Persönlich-
keiten der Wiener Medizinischen Schule wie Ernst Brücke und Theodor
Billroth wesentlich bereichert und ergänzt. Mit Nietzsche kam es im Winter
1883/84 während eines mehrmonatigen Aufenthalts in Nizza zu einem
intensiven geistigen Austausch, der in zahlreichen Briefen umfassend doku-
mentiert ist und hier erstmals vollständig nach den Handschriften zugäng-
lich gemacht wird. Darüberhinaus dokumentieren die Breife die zeitgenös-
sische Rezeption Kants und Schopenhauers und stellen ausführlich den
Stand der psychologischen Diskussion über das Wesen des Bewußtseins und
das Unbewußte dar. Die Frage nach der Möglichkeit einer physikalischen
Psychologie beantwortet Paneth in seinem Aufsatz "Die Erhaltung der
Energie auf psychischem Gebiete", "Quid faciendum", der zweite Essai,
liefert eine eigenwillig tiefgründige Analyse der Rolle des Judentums kurz
vor dem Aufkommen des Zionismus. Ein Sterbebericht Paneths aus der
Hand seiner Frau Sophie, ein Brief Sophies an Freud sowie ein Brief
Nietzsches ergänzen die ausführlich eingeleitete, wissenschaftlich kommen-
tierte Ausgabe, deren Titel auf das Herzstück des Bandes hindeutet: Josef
Paneths Autobiographie Vita nuova.

USA/Canada: Editions Rodopi, 233 Peachtree Street, N.E., Suite 404, Atlanta, Ga. 30303-
1504, Telephone (404) 523-1964, only USA 1-800-225-3998, Fax (404) —522-7116
And Others: Editions Rodopi B.V., Keizersgracht 302-304, 1016 EX Amsterdam, The
Netherlands. Telephone (020) — 622.75.07, Fax (020) — 638.09.48

TRACTRIX

Yearbook for the History of Science, Medicine, Technology & Mathematics
Vol. 2 1990. Amsterdam/ Atlanta GA. 167 pp. ISSN: 0924-0829.
Subscriptions: The price of *Tractrix* is Hfl. 67,—/US-$ 33.50 for institutions, and Hfl. 35,—/US-$ 17.50 for individual subscribers who place their order directly with the publishers (postage Hfl. 7,—/US-$ 3.50).

Contents. ARTICLES. Marian Fournier: The book of nature: Jan Swammerdam's microscopical investigations. Kirsti Andersen: Stevin's theory of perspective: the origin of a Dutch academic approach to perspective. Harold J. Cook: Medical innovation or medical malpractice? Or, a Dutch physician in London: Joannes Groenevelt, 1694-1700. Marta Fehér: The triumphal march of a paradigm: a case study of the popularization of Newtonian science. DUTCH CLASSICS. E.J. Dijksterhuis: Ad quanta intelligenda condita (Designed for grasping quantities). WORK IN PROGRESS. Dick van Lente: Technology in Dutch society during the nineteenth century. Bert Theunissen and Robert P.W. Visser: History of biology in the Netherlands: an historical sketch. ESSAY REVIEWS. H.F. Cohen: 'Open and Wide, yet Without Height or Depth'. Review of: K. van Berkel, In het voetspoor van Stevin. Geschiedenis van de natuurwetenschap in Nederland 1580-1940 ('In Stevin's Footsteps. A History of Science in the Netherlands 1580-1940'), (Meppel: Boom, 1985; ISBN 90-6009-639-8), 243 pp., ill. Dissertations completed in the Netherlands in 1989/1990.

CLIO MEDICA

Vol. 22. Essays in the history of therapeutics. Edited by W.F. Bynum and V. Nutton. Amsterdam/Atlanta, GA 1991. ca. 200 pp.
ISSN: 0045-7183   ca. Hfl. 80,—/US-$ 40.—
Contents: The Editors: Introduction. Guenter B. Risse: The History of Therapeutics. Vivian Nutton: From Medical Certainty to Medical Amulets: Three Aspects of Ancient Therapeutics. Luis Garcia-Ballester: Dietetic and Pharmacological Therapy: A Dilemma among Fourteenth-Century Jewish Practitioners in the Monpelier Area. Christa Habrich: Characteristic Features of Eighteenth-Century Therapeutics in Germany. Renate Wittern: The Origins of Homoeopathy in Germany. John Harley Warner: Science, Healing and the Physician's Identity: A Problem of Professional Character in Nineteenth-Century America. Ulrich Tröhler: 'To Operate or not to Operate?' Argumentation in Therapeutical Controversies within the Swiss Society of Surgery 1913-1988.

USA/Canada: Editions Rodopi, 233 Peachtree Street, N.E., Suite 404, Atlanta, Ga. 30303-1504, Telephone (404) 523-1964, only USA 1-800-225-3998, Fax (404)—522-7116
And Others: Editions Rodopi B.V., Keizersgracht 302-304, 1016 EX Amsterdam, Telephone (020) —22.75.07, Fax (020) — 38.09.48

GRAZER PHILOSOPHISCHE STUDIEN

Vol. 35.   Hfl. 75,—/US-$ 37.50
Contents: John Haldane: Brentano's Problem. Dieter Münch: Brentano and
Comte. James Petrik: Two Faces Have "I". Monica Holland: Emotional as a
Basis of Belief. Andrew Ward: The Relational Character of Belief. Uwe
Meixner: Descartes' Argument für den psycho-physischen Dualismus im
Lichte der modal-epistemischen Logik. Arno Ros: "Begriff", "Setzung",
"Existenz" bei W.V.O. Quine. Rainer W. Trapp: Systematische Klassifika-
tion und vergleichende Betrachtung der wichtigsten Ethiktypen unter dem
Gesichtspunkt ihrer Eignung als allgemein akzeptable Handlungsrichtlinien.
Review Articles. Critical Notes.

Vol. 36. **The Mind of Donald Davidson.** Edited by Johannes Brandl and
Wolfgang L. Gombocz. Amsterdam/Atlanta GA 1989. 219 pp.
ISBN: 90-5183-137-4.   Hfl. 60,—/US-$ 30.—
Contents: Donald Davidson: What is Present to the Mind? Ullin T. Place:
Thirty Five Years On — Is Consciousness Still a Brain Process? Peter Lanz:
Davidson on Explaining Intentional Actions. Matthias Varga von Kibéd:
Some Remarks on Davidson's Theory of Truth. Ernest Lepore and Barry
Loewer: What Davidson Should Have Said. Johannes Brandl: What is
Wrong with the Building Block Theory of Language? Eva Picardi: Davidson
on Assertion, Convention and Belief. Hans Georg Zilian: Convention and
Assertion. Dunja Jutronič-Tihomirovič: Davidson on Convention. Damjan
Bojadžiev: Davidson's Semantics and Computational Understanding of
Language. Joachim Schulte: Wittgenstein's Notion of Secondary Meaning
and Davidson's Account of Metaphor — A Comparison. Arto Siitonen:
Understanding Our Actual Scheme. J.E. Malpas: Ontological Relativity in
Quine and Davidson. Matjaž Potrč: Externalising Content. Donald David-
son: The Conditions of Thought.

Individual subscribers who directly place their orders with the publishers:
50% discount (postage Hfl. 5,—/US-$ 2.50)

USA/Canada: Editions Rodopi, 233 Peachtree Street, N.E., Suite 404, Atlanta, Ga. 30303-
    1504, Telephone (404) 523-1964, only USA 1-800-225-3998, Fax (404) — 522-7116
And Others: Editions Rodopi B.V., Keizersgracht 302-304, 1016 EX Amsterdam,
    Telephone (020) —22.75.07, Fax (020) — 38.09.48

GRAZER PHILOSOPHISCHE STUDIEN

Internationale Zeitschrift
für analytische Philosophie

International Journal for
Analytic Philosophy

Herausgeber   Rudolf Haller   Editor

BAND 38   1990   VOLUME 38

Grazer Philosophische Studien is a philosophical Journal publishing papers in
Analytic Philosophy generally and in Austrian Philosophy especially.
Herausgeber/Editor: Prof. Dr. Rudolf Haller, Institut für Philosophie,
Universität Graz, Heinrichstraße 26, 8010 Graz, Österreich/Austria.

Verlag/Publisher:
USA/Canada: Editions Rodopi, 233 Peachtree Street, N.E., Suite 404, Atlanta, Ga. 30303-
     1504, Telephone (404) 523-1964, only USA 1-800-225-3998, Fax (404) — 522-7116
And Others: Editions Rodopi B.V., Keizersgracht 302-304, 1016 EX Amsterdam,
     Telephone (020) —22.75.07, Fax (020) — 38.09.48

MORAL UND POLITIK
AUS DER SICHT DES KRITISCHEN RATIONALISMUS

Hrsg. v. Kurt Salamun. Amsterdam/Atlanta GA 1990. ca. 300 pp. (Schriften-
reihe zur Philosophie Karl R. Poppers und des Kritischen Rationalismus 1).
ISBN: 90-5183-203-6   Bound ca. Hfl. 120,00/US-$ 60.00
ISBN: 90-5183-204-4   Paper ca. Hfl. 49,50/US-$ 24.75

Inhalt: I. *Freiheit, Moral und Ethos der Aufklärung.* Hans Albert: Die
Verfassung der Freiheit. Bedingungen der Möglichkeit sozialer Ordnung.
Volker Gadenne: Karl Poppers Beitrag zum Problem der Willensfreiheit.
Yoshihisha Hagiwara: Zum Verständnis von Liberalismus bei Popper und
Hayek. Klaus Mühlfeld: Bemerkungen zur binären Struktur der Moral. Kurt
Salamun: Das Ethos der Aufklärung im Kritischen Rationalismus.
II: *Zur Idee der offenen Gesellschaft.* Evelyn Gröbl-Steinbach: Von der
offenen zur postmodernen Gesellschaft? Hardy Bouillon: Politische Philo-
sophie im Rahmen einer offenen Gesellschaft: Anmerkungen zu Popper und
Hayek. Dariusz Aleksandrowicz: Die Krämer und die Ritter. Vom Ethos der
offenen Gesellschaft, der Teilung Europas und der Politik des Friedens.
Lothar Schäfer: Kritischer Rationalismus und ökologische Krise: Überlegungen zur Utopie- und Technikkritik.
III: *Demokratie, Souveränität und soziale Marktwirtschaft.* Fred Eidlin:
Popper und die demokratische Theorie. Andras Pickl: Fallibilismus und die
Grundprobleme der Politischen Theorie: Zu Poppers Kritik der Souveräni-
tätstheorie und seinem Neuansatz für die politische Theorie. Gerard Rad-
nitzky: Die politische Philosophie des Kritischen Rationalismus und die
soziale Marktwirtschaft.

USA/Canada: Editions Rodopi, 233 Peachtree Street, N.E., Suite 404, Atlanta, Ga. 30303-
1504, Telephone (404) 523-1964, only USA 1-800-225-3998, Fax (404) — 522-7116
And Others: Editions Rodopi B.V., Keizersgracht 302-304, 1016 EX Amsterdam,
Telephone (020) —22.75.07, Fax (020) — 38.09.48

Printed in the United States
by Baker & Taylor Publisher Services

Printed in the United States
by Baker & Taylor Publisher Services